YOUR KNOWLEDGE HAS VALUE

- We will publish your bachelor's and master's thesis, essays and papers

- Your own eBook and book - sold worldwide in all relevant shops

- Earn money with each sale

Upload your text at www.GRIN.com and publish for free

Syed Arshad Hussain

Langmuir-Blodgett Films

GRIN Verlag

Imprint:

Copyright © 2010 GRIN Verlag GmbH
Druck und Bindung: Books on Demand GmbH, Norderstedt Germany
ISBN: 978-3-656-08944-5

LANGMUIR-BLODGETT (LB) FILMS

Syed Arshad Hussain*

Department of Physics, Tripura University,

The Langmuir-Blodgett (LB) technique is a way of making ultrathin nanostructured films with a controlled layer structure and crystal parameter, which have many envisioned applications in technology for optical and molecular electronic devices as well as in signal processing and transformation. LB films have a unique potential for controlling the structure of organized matter on the ultimate scale of miniaturization, and must surely find a niche where this potential is fulfilled. In this review paper the early development, technology and present status of LB film research and key papers are referred to.
Key words: Langmuir-Blodgett films, thin film, amphiphilic and π-A isotherm.
Pacs No: 78.20.-e, 81.40,42.65,79.60.Dp, 79.60.Jv

1. Introduction:

Ultrathin films fabricated by Langmuir-Blodgett (LB) technique is a powerful tool in creating carefully controlled supramolecular structures of organized molecular assemblies, which have their potential applications in optical, electrical and numerous other fields. The bulk properties of the materials incorporated into Langmuir-Blodgett (LB) films can be controlled by the organization of molecules in the films and also by changing various LB parameters. Thin films with carefully designed structures and properties have attracted a great deal of interest in recent years [1] due to their potential applications in a number of different fields, such as sensors,[2] detectors,[1] surface coating,[1] optical signal processing,[3] digital optical switching devices,[4] molecular electronic devices,[5] nonlinear optics and models mimicking biological membranes.[6] These applications require, in general, well ordered films consisting of molecules with specific properties, carefully aligned with respect to each other and the substrates possessing high degree of stability to thermal and chemical changes. The possibility to synthesize organic molecules almost without limitations and with desired structure and functionality in conjunction with Langmuir-Blodgett (LB) film deposition technique enables the production of electrically, optically and biologically active components on a nanometer scale. It is extremely important to get an idea about the relation between the ultrastructure as well as also the domain structure and various physical properties of such system. The basic physics involved in such characterizations is a topic of fundamental importance. The structure, morphology, configuration and various other physical properties of the films obtained by Langmuir-Blodgett (LB) technique can be easily modified, that may suit for any specific application.

2. History and Development:

The interesting effects of oil on water were known to the ancients. According to Tabor,[7] the earliest account is to be found in cuneiform on clay tablets prepared during Hammurabi's period in the eighteenth century BC; this describes how the Babylonians used to practice divinity by observing the spreading behaviour of oil on water. The earliest technical application of organic monolayer films is believed to be the Japanese printing art called sumi-nagashi. The dye comprising a suspension of submicron carbon particles and protein molecules is first spread on the surface of water; the application of gelatin to the uniform layer converts the film into a patch-work of colourless and dark domains.[8] These distinctive patterns can then be transferred by lowering a sheet of paper onto the water surface. However, the first attempt to place the subject on a scientific basis was made by Benjamin Franklin, the versatile American statesman, in the eighteenth century AD. It was during his frequent visits to Europe as the principal representative of the American States in their critical discussions on sovereignty with the French and the British that he carried out his original experiments on the spreading of oil on various stretches of English water. His famous communication [9] to the Royal Society in 1774, reporting how a teaspoonful of oil had a calming influence over half an acre of a pond on

Clapham Common, has been described as the first recorded scientific experiment in surface chemistry. Franklin also speculated, as Plutarch, Aristotle and Pliny had done,[10] on the practical utility of oil spreading for calming rough seas. In fact, at about this time, an enterprising Scotsman, John Shields [11] lodged a patent for spreading oil from valves in underwater pipes at the entrance to harbours, and a supporter of the principle, Miss C. F. Gordon-Cumming [12] even suggested that oil bags be provided on each lifebuoy! Despite some parliamentary interest there was little follow up and indeed one potential sponsor declined support because the experiments "would require an expenditure of time and funds not at present at our disposal"!

Franklin must have been too preoccupied with political affairs to place his observations of oil on water on a simple quantitative basis. Had he done so, he might well have calculated that a volume of one teaspoonful (~ *2ml)* spread over an area of nearly half an acre (,~2000m^2) leads to a surface coating approximately 1 nm thick. Lord Rayleigh [13] is given the distinction of first suspecting that the maximum extension of an oil film on water represented a layer one molecule thick. He had been convinced since 1865 of the reality of molecules but his early experiments were not definitive and involved estimating the amount of oil capable of preventing the movement of camphor on water. For a direct measurement of molecular sizes he was indebted to Agnes Pockels [14] whose simple apparatus later became the model for what is now termed a Langmuir trough. Rayleigh confirmed the results of many of Pockels' experiments, which had been carried out essentially in a kitchen sink. He reported [15] in 1899 the precise thickness of a monomolecular layer of castor oil on water to be 1 nm. However, this significant observation went unrecognised until monolayer theory was more fully developed by Langmuir in the General Electric Company laboratories at Schenectady, New York between the two World Wars. Langmuir appreciated the applicability of the monolayer concept to the adsorption of gases or solutes by solids. The experiments for which he was awarded his Nobel Prize indicated that adsorbed films of very great stability were formed in which single layers of atoms were bound by the underlying surface. In order to test the general applicability of his hypothesis about the involvement of short range forces he turned his attention to liquids. In one of his earliest papers [16] he described his film balance or trough and showed, as Pockels had done earlier, that it could be utilized to elucidate the shapes and sizes of molecules and their orientation at the air-liquid interface. Most of his experiments were on a well defined series of fatty acid salts. As early as 1920 [17] he reported the transfer of such molecules from a water surface onto a solid support. However, the first detailed description of sequential monolayer transfer was given several years later by Blodgett.[18] These built up monolayer assemblies are now called Langmuir-Blodgett (LB) films; the floating monolayer is usually referred to as a Langmuir film. The publications resulting from the pioneering experiments carried out by these two investigators during the period 1934 and 1952 have been summarized by Gaines [19] whose book [20] provides a comprehensive treatment of the subject up to 1966.

After the pioneering work done by Langmuir and Blodgett, it took almost half a century before scientists all around the world started to realize the opportunities of this unique technique. In the late 1960's Hans Kuhn described Langmuir-Blodgett (LB) films with designated properties.[21] Extensive studies on LB films are being conducted after the work of Kuhn and co-workers.

The first international conference on Langmuir-Blodgett (LB) films was held in 1979 and since then the use of this technique is increasing among the scientists throughout the world working on various different field of research.

To illustrate the global effort on thin film research, a graphic was made based on a single data bank internet research (Figure 1). It is the total number of papers containing monolayer in the title. It cauld be seen that there were a relatively slow start in the 60's, but the interest exploded in the 70's. The trend was linear for 20 years up to the middle of the 90's. The number of publications reached a maximum of 1322 in 1998. A total of 27 226 reports were published according to that data bank.

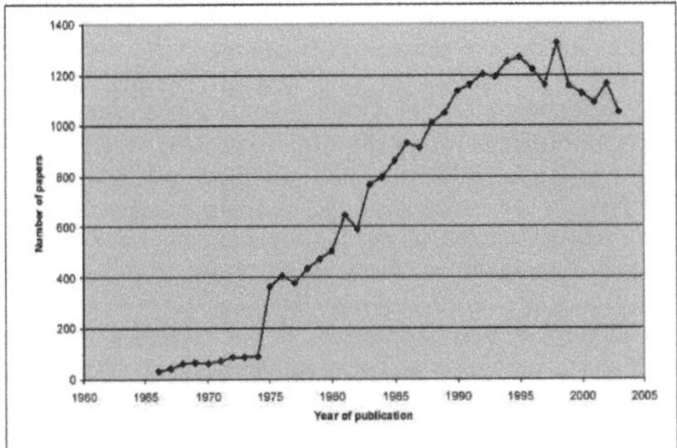

Fig. 1 Number of publications per year with "monolayer" in the title.

3. Various Terms related to Langmuir-Blodgett (LB) Films:

3.1 LB Compatible Materials:

Molecules consist of a hydrophilic (water soluble) and a hydrophobic (water insoluble) part are ideal for Langmuir-Blodgett (LB) film deposition. This amphiphilic nature of these molecules is responsible for their association behaviour in solution and their accumulation at interfaces (air/water). The hydrophobic part usually consists of hydrocarbon or fluorocarbon chains, while the hydrophilic part consists of a polar group (-OH, -COOH, $-NH_3^+$, $-PO_4-$ $(CH_2)_2NH_3^+$ etc.). These types of molecules are known as amphiphilic molecules.

Surfactants are good example of amphiphilic molecules possessing a hydrophilic and a hydrophobic part. Complex association behaviour of surfactants is observed in solution and they accumulate at the air-water interface due to the 'amphiphatic balance' i.e. the balance between hydrophilic and hydrophobic constituents within the same molecule. Depending on various parameters surfactants are observed to form different interesting structures at the air-water interface viz, micelles, bilayer, vesicles, aggregates, monomers, dimers, n-mers etc.

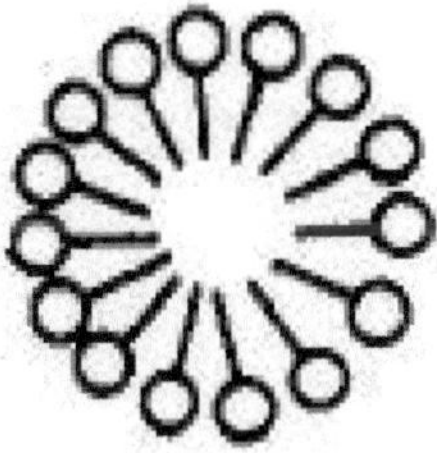
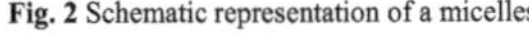
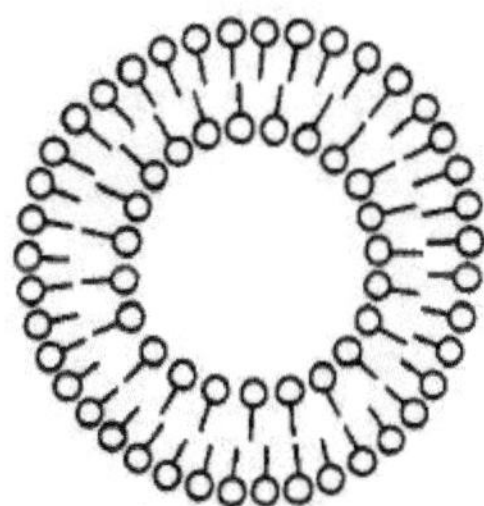

Fig. 2 Schematic representation of a micelles   **Fig. 3** Schematic representation of a vesicle.

Micelles consist of aggregates of amphiphilic molecules arranged so that, in polar solvent, the polar groups face outwards and the non-polar groups are bounced together at the centre of the aggregate (figure 2). For many systems there is a critical concentration at which such structures start to form and this is known as the critical miceller concentration (CMC). In a non polar solvent such structure can also exist but with the polar groups facing inwards and the non-polar groups facing outwards.

Vesicles consist of closed structures usually, approximately spherical in form, whose walls are made from bilayers of amphiphilic molecules (figure 3). They both reside in and contain solvent.

In general, materials such as long chain fatty acids and alcohols are used as film forming materials. The hydrocarbon part $-CH_2$ of the molecule is responsible for the material's repulsion of water while the polar -COOH or -OH group has sufficient affinity for water to anchor the molecule in the aqueous subphase. At the air-water interface such amphiphilic molecules orient themselves with their hydrocarbon chains (hydrophobic part) upwards in the air and their hydrophilic head groups tend downwards into the water.

The materials widely studied are the alkanoic acids and their salts and long chain alcohols. The length of the hydrocarbon chain is critical and its effect on the surface tension of the subphase is given by Traube's rule, an approximation which states that for a given homologous series of amphiphiles the concentration required for an equal lowering of surface tension in dilute solution decreases by a factor of about three for each additional CH_2 group.

Although traditionally long hydrocarbon chains have been considered necessary, the incorporation of aromatic moieties and unsaturated chains into the amphiphile tends to decrease the hydrocarbon chain length necessary for deposition. For example, a substituted anthracene derivative has been successfully deposited with only a butyl chain attached to the polyaromatic system. In fact, with appropriate synthetic chemistry almost any active species can be tailored into a form which would float as a monolayer.

Typical examples of LB compatible molecules are Stearic acid (SA), Arachidic acid (AA) and Zinc (Arachidate)$_2$ etc. They are shown in the figures 4 and 5. Generally these molecules having interesting physical properties (viz. conducting, semi conducting, lasing action, gas sensing, pyro- electric etc.), are attached to a long hydrophobic chain (alkyl chain) to make LB compatible. Other ways to make a molecule LB compatible is to attach the molecule into a polymeric backbone

However both these processes require expertization in synthesis and also time consuming and quite costly. Moreover purifications of these synthesized materials are quite laborious and need a lot of expertization.

Recently it was observed that certain non-amphiphilic molecules (i.e, without any hydrophobic chain) when mixed with a long chain fatty acid (viz. stearic acid or arachidic acid) or with an inert polymer matrix (polymethyl methacrylate or polystyrene) form excellent LB films.[1]

Actually in these cases long chain fatty acid or polymer act as building materials which support the non-amphiphilic molecules to organize at the air water interface.

Since these non-amphiphilic molecules are easily available and they have wide range of varieties having different interesting physical properties hence a lot of investigations on LB films using these molecules are important. It was also observed that the quality of LB films of these non-amphiphilic molecules is comparable to their amphiphilic counterparts with respect to their spectroscopic and aggregating properties.

Another important aspect of these types of mixed LB films is that in some cases the optical and electrical characteristics of materials are changed markedly when they are incorporated into some inert matrices of restricted geometries. Detailed investigations of these films are extremely important from the point of view of their various technical applications.

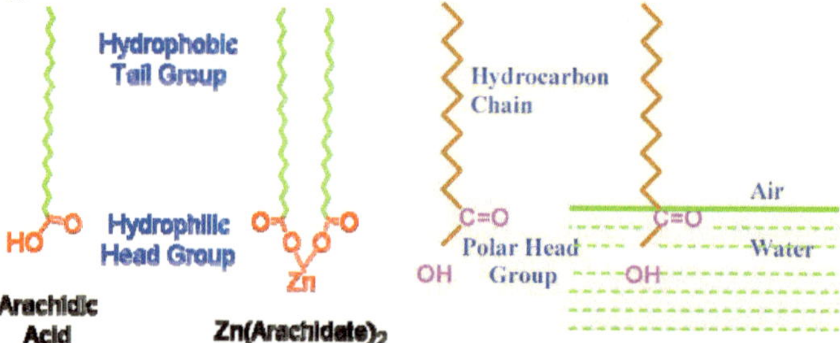

Fig. 4 Ideal LB compatible molecule

Fig. 5 The components of an ideal amphiphilic and its orientation adopted on the subphase.

The association behaviour of surfactants in solution and their affinity for interfaces are determined by the physical and chemical properties of the hydrophobic and hydrophilic groups, respectively.

3.2. Surface Pressure:

The air-water interface possesses an excess free energy originating from the differences in environment between the surface molecules and those in the bulk. This interfacial free energy can be determined by measuring the Surface Tension. Surface Tension can be defined as the work required for expanding the surface isothermally by unit area and is quantified as the force/length measurement. It is also said to be the measurement of the cohesive energy present at an interface. At room temperature the Surface Tension of pure water is around 70 mN/m, which is an exceptionally high value compared to other liquids and consequently makes water a very good subphase for monolayer studies.

When some impurity is added on the water surface then the Surface Tension is reduced. Also when a monolayer is compressed on the water surface, the instantaneous Surface Tension of that surface is reduced due to increase in impurity concentration on the surface. The Surface Tension of the pure water is taken as the zero level reference and the decrease in Surface Tension can be measured with respect to this reference level and gives negative value. Conventionally in LB technique, this negative or decrease in Surface Tension is referred to as an increase in positive Surface Pressure. Numerically both the Surface Tension and the Surface Pressure are same. However the Surface Pressure is represented along the positive Y-axis unlike the Surface Tension, which is along the negative Y-axis. Surface Tension and Surface Pressure have the same unit and magnitude but Surface Pressure increases as the Surface Tension decreases.

If the available surface area of the monolayer is reduced by a barrier system the molecules start to exert a repulsive effect on each other. This two-dimensional analogue of a pressure is called Surface Pressure, and is given by the following relationship,

$$\pi = \gamma - \gamma_0$$

where γ is the Surface Tension in the absence of a monolayer and γ_0 is the Surface Tension with the monolayer present.

3.3. Surface pressure vs. area per Molecule ($\pi - A$) isotherm and formation of Langmuir monolayer at the air-water interface:

The most important indicator of the monolayer properties of an amphiphilic material is given by measuring the surface pressure as a function of the area of water surface available to each molecule. This is carried out at constant temperature and is known as a surface pressure vs. area per molecule isotherm ($\pi - A$) or simply "isotherm". Usually an isotherm is recorded by compressing the film (reducing the area with the barrier) at a constant rate while continuously monitoring the surface pressure (π). Depending on the material being studied, repeated compressions and expansions may be necessary to achieve a reproducible trace. A schematic representation of surface pressure vs. area per molecule isotherm is shown in figure 6. The pressure-area ($\pi - A$) isotherm is rich in information on the stability of the monolayer at the air-water interface, the reorientation of molecules in the two dimensional system, phase transitions and conformational transformations.

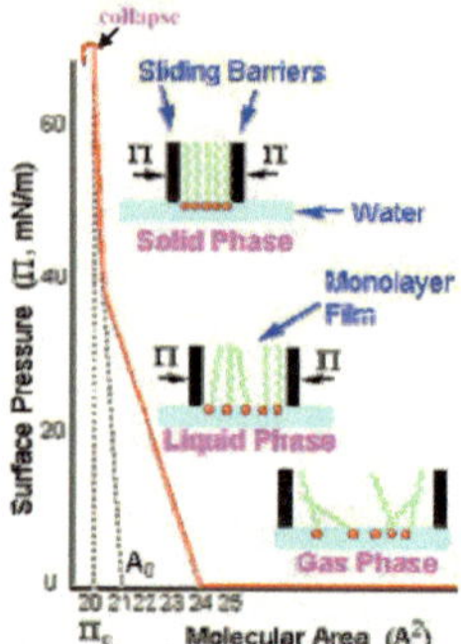

Fig. 6 Schematic representation of pressure-area isotherm.

First of all a minute amount of suitable LB compatible material is dissolved in a highly volatile and water insoluble medium (viz. chloroform) to prepare a dilute solution. Now a few micro liters of this solution is spread at the air-water interface (i.e., on the water surface) very slowly by means of a micro syringe. The amount of the molecule should be calculated before spreading so that the resulting film would be a monomolecular layer. Now at this stage after allowing sufficient time (15-20 minute) to evaporate the solvent, the barrier is compressed very slowly and the corresponding pressure and area per molecule are recorded. The plot of surface pressure (π) as a function of area per molecule (A) is known as the pressure-area ($\pi - A$) isotherm. A schematic representation of pressure-area isotherm is shown in figure 6. A number of distinct regions is immediately apparent on examining the isotherm. These regions are called phases. As one can see when the monolayer is compressed it can pass through several different phases, which are identified as discontinuities in the isotherm.

The phase behaviour of the monolayer is mainly determined by the physical and chemical properties of the amphiphile, the subphase temperature and the subphase composition. For example, various monolayer states may exist depending on the length of the hydrocarbon chain length and the magnitude of other cohesive and repulsive forces existing between head groups. An increase in the chain length increases the attraction between molecules, condensing the π–A isotherm. On the other hand, if an ionisable amphiphile is used, the ionization of the head groups induces repulsive forces tending to oppose phase transitions.

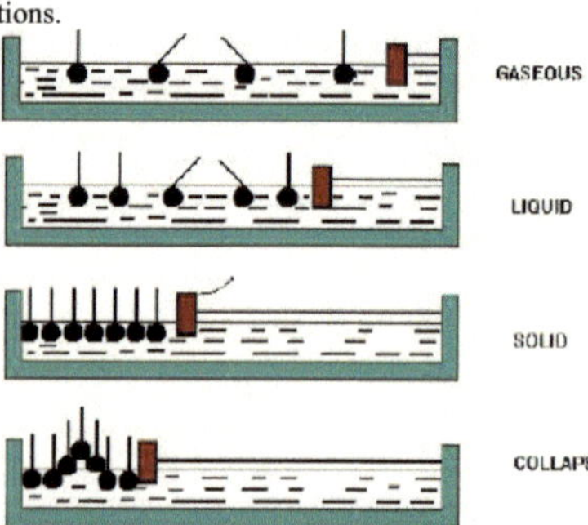

Fig. 7 Schematic representation of molecules at different phases during monolayer compression.

Initially the spread molecules remain on the water surface haphazardly and move randomly on the two-dimensional water surface as if like gas molecules in three-dimensional space. This situation of the molecules on the two- dimensional water surface is termed as gas phase and in this phase the molecules are very loosely packed. In this gas phase, each molecule occupies a large molecular area (A) and the surface pressure (π) is quite low as shown in figure 6 and 7. The area per molecule in the gaseous- phase is large and ideally there should be no interaction (lateral adhesion) between the molecules and hence the surface pressure is very low. The surfactant molecules have natural tendency to aggregate, even when the surface pressure approaches zero. For an ideal two dimensional gaseous phase, the molecules will obey the two dimensional ideal gas equation $\pi A = KT$, where A is the area per molecule, π is the surface pressure and T is the absolute temperature. It is interesting to note that A is equivalent to V and π is equivalent to P in the ideal gas law. Now as the molecules are compressed further by a moveable barrier, the molecules get closer resulting in the decrease in intermolecular distance and increase in surface pressure. Also the phase transitions in isotherm are observed. The first change in compressibility of the $\pi - A$ isotherm curve signifies the onset of 'liquid-expanded' state. Here the molecules are not as free as to move about as in the gas phase. Upon further compression, the monolayer may enter into the liquid-condensed state. Here the molecules occupy a smaller area and lower degrees of freedom than in the liquid-expanded phase. Further compression of the film may causes a second phase transition, from liquid-condensed to solid state. This solid state is characterized by the steep and usually linear relationship between the surface pressure and molecular area. Here the molecules are closely packed and with high density and uniformly oriented. At this stage the area per molecule for stearic acid is ≈ 0.21 nm^2. If the monolayer is further compressed it collapses due to mechanical instability.[22] The collapse is generally seen as a rapid decrease in the surface pressure or as a horizontal break in the isotherm.[23] This collapse pressure (π_c) is a function of temperature, pH of the subphase and the speed of the barrier compression. It is also observed that there are certain materials that exhibit two instances of collapse pressure. Polypeptides show such characteristics,[23] they first collapse to form bilayers and then again, to form microcrystallites.[23]

When the monolayer is in the two-dimensional 'solid' or 'liquid condensed' phase the molecules are relatively well oriented and closely packed and the zero-pressure molecular area (A_0) can be obtained by extrapolating the slope of the 'solid' phase to zero pressure - the point at which this line touches the x-axis is the hypothetical area occupied by one molecule in the condensed phase at zero surface pressure (Figure 6). For a fully saturated alkanoic acid such as stearic acid, the molecular area, determined in this way is 22-25 nm^2. Such an area corresponds to the cross-sectional area of a hydrocarbon chain, suggesting that the compressed monolayer consists of close packed vertically orientated chains.

From the π-A isotherm study quantitative information can be obtained on the molecular dimensions and shape of the molecules under study. It is important to study the isotherms of the monolayer of the film forming materials as a pre-requisite to the determination of the dipping characteristics. Information can be obtained as to the way in which the molecules pack at the air-water interface and the stability of the compressed layer at high surface pressure.

In general $\pi - A$ isotherm represents the fingerprint to the given molecule. The isotherms are strongly sensitive to the nature of the molecules, its orientation, compositions of mixture etc and also there is a strong correlation between the molecular shape and the isotherm characteristics.[24]

3.4. Forces Responsible for the Formation of Monolayer and Its Stability:

Surfactant molecules are amphiphilic in nature and are composed of a hydrophilic part and a hydrophobic part. Hydrophilic groups are consist of polar groups such as carboxylic acid, sulphates, amines, alcohols etc. These are all attracted to the polar media such as water and the forces acting on them are predominantly ionic and coulombic type ($1/r^2$), r being the intermolecular separation). On the other hand hydrophobic groups such as a hydrocarbon chain, fats and lipids etc are almost water insoluble and the forces acting on them are predominantly Van der Waal's $(1/r^{12} \, or \, 1/r^6 \quad 1/r^{12} \, or \, 1/r^6)$. The sum of these attractive and repulsive forces as well as the change in entropy during monolayer compression plays important role in determining the phase behaviour of the monolayer at the air-water interface. The origin of these forces has been discussed in detail by Israelachvile.[25] However the magnitude of the individual contribution of the above mentioned forces is governed by the temperature at which the experiment is carried out and also the molecular structure of the surfactant. The pH and composition of the subphase will also influence the phase behaviour of the spread monolayer. Generally an increase in temperature causes an expansion of monolayer due to the increase in thermal movement of the constituent molecules.

3.4.1. Electrostatic interaction:

The forces affecting the polar head groups submerged in the aqueous subphase are ionic and include mainly dipole-dipole interactions. However if the constituent molecules of the floating monolayer are partly ionized, ion-ion and ion-dipole interaction may take place.[26] The ion-dipole and ion-ion interactions are inversely proportional to the second and third power of the intermolecular separations respectively. Increase in ionization will lead to an expansion of the monolayer due to the repulsion between the ions of the same kind (same sign).

However, the dissociation in the bulk is energetically different in comparison to the dissociation at the interface.

3.4.2. Van der Waals force:

The weak Van der Waal's attractive forces between the hydrophobic part are mainly responsible for the condensed states in the monolayer. These forces are mainly short range and inversely proportional to the sixth power of the intermolecular separation. However depending on the quantum mechanical calculations for long, linear hydrocarbon chains a fifth order dependence has been suggested.[27] Again for saturated hydrocarbon the extent of the force is directly proportional to the number of carbon atoms present in the carbon chain. The presence of bulky constituents like methyl group in the hydrocarbon chain or if the packing of the hydrocarbon chain is hindered by the introduction of the double bonds, causes a reduction in the cohesive forces between the constituent molecules.

There are several factors affecting the monolayer stability. Evaporation of the molecules,[28] dissociation into the subphase,[29] reorganization and collapse into three dimensional bulk crystals (micro-crystal or nano-crystal) and aggregate formation [30] are observed to cause instability of the monolayer. Changes in subphase composition may also affect the monolayer stability.

3.5. Transfer of the Langmuir monolayer onto solid substrate: formation of Langmuir-Blodgett (LB) films:

Apart from being used as a tool for monolayer investigation, the most celebrated features of the LB technique is the facility of producing mono- and multilayer of controlled thickness of these films onto solid substrates.[31] These are highly organized and defect free films of molecular dimension. In general the transfer of Langmuir monolayer from the air-water interface onto a solid substrate can be made by successfully dipping a solid substrate up and down through the spread monolayer while simultaneously keeping the surface pressure fixed by a computer controlled feedback system between the elctrobalance measuring the surface pressure and the barrier moving mechanism. K. Blodgett and I. Langmuir first demonstrated this technique [32] and accordingly this is called Langmuir-Blodgett (LB) technique. The film obtained by this technique is called the Langmuir-Blodgett (LB) films.

The Langmuir-Blodgett (LB) film deposition is traditionally carried out in the "solid" phase. The surface pressure is then high enough to ensure sufficient cohesion in the monolayer, e.g. the attraction between the molecules in the monolayer is high enough so that the monolayer does not fall apart during transfer to the solid substrate. This also ensures the build up of homogeneous multilayer. The surface pressure value that gives the best result depends on the nature of the monolayer and is usually established empirically.

When a substrate is moved through the monolayer at the air-water interface, the first monolayer is transferred onto the substrate during upstroke or down stroke respectively depending on the nature of the substrates. Although there are many different types of substrate materials are possible, however they can be divided into two different categories with distinctly different behaviour depending on the deposition of the first layer. For hydrophilic substrates (a category which includes all metals and semi-conductors with a native oxide layer and glass as normally prepared) no monolayer is deposited as the substrate is initially immersed, so that the eventual number of layers when the substrate finally emerges into the air is odd. On the other hand hydrophobic substrates (including silicon and other semiconductors when oxide-free, and silanised glass) take up a monolayer on the first down stroke as well so that the final number of layers is even.

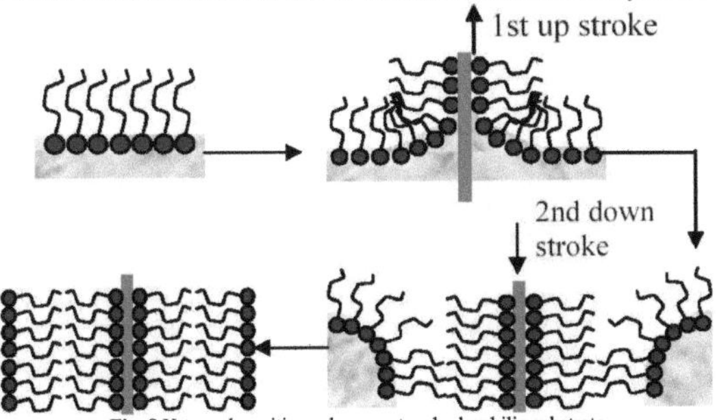

Fig. 8 Y-type deposition scheme onto a hydrophilic substrate

Depending on the nature of the spread monolayer and the surface of the substrate different LB film structures are observed. A monolayer will be transferred during upstroke when the substrate is hydrophilic and the hydrophilic

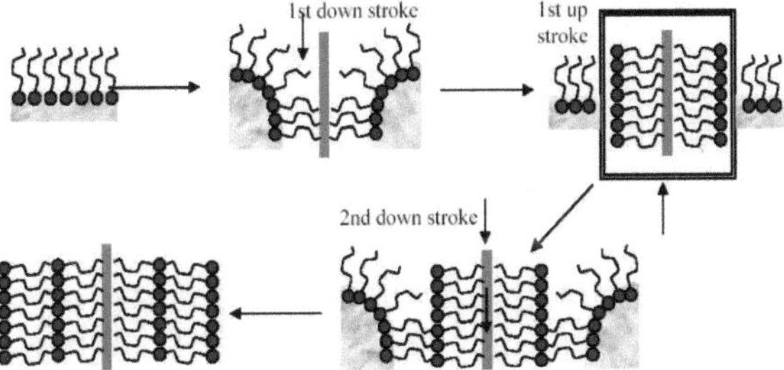

Fig. 9 X-type deposition schemes.

head group interacts with the surface of the substrate. On the other hand, if the substrate is hydrophobic, the monolayer will be transferred during first down stroke and the hydrophobic tail group interacts with the substrate surface. A hydrophilic substrate becomes hydrophobic after first monolayer transfer and thus the second layer will be transferred during the down stroke as shown in figure 8. This mode of film deposition is called Y-type deposition. In this way multilayers of LB films can be prepared on a solid substrate as shown in figure 11. This is the most stable film deposition procedure as the interaction between the adjacent monolayers are similar i.e., hydrophobic-hydrophobic or hydrophilic-hydrophilic in natureHowever in some cases, the deposition occurs during the down or upstrokes only resulting a head to tail or tail to head arrangement of multilayers and the corresponding modes of depositions are called X-type or Z-type respectively (figures 9 and 10).

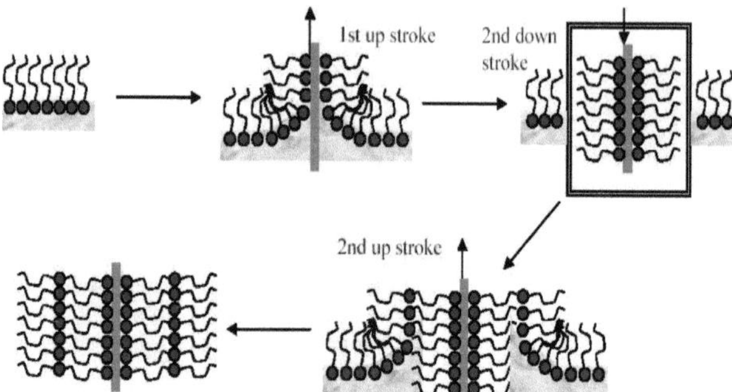

Fig. 10 Z-type deposition scheme.

Here in both cases the films are less stable compared to Y-type as the interaction between the adjacent monolayer is hydrophobic-hydrophilic. Both the X- type and Z-type films may have importance in non-linear optical applications due to their non-centrosymmetric structure.[6] Intermediate structures are sometimes observed for some LB multilayers and they are often referred to be XY-type multilayers.[33] The production of alternating layers which consist of two different kinds of amphiphiles is also possible by using highly sophisticated double trough instruments.

The quantity and the quality of the deposited monolayer on a solid substrate is measured by transfer ratio, t_r. This is defined as the ratio between the decrease in monolayer area during a deposition stroke, A_d, and the area of the substrate, A_s i.e.

$$t_r = \frac{A_d}{A_s}$$. For ideal transfer the t_r. is equal to 1.

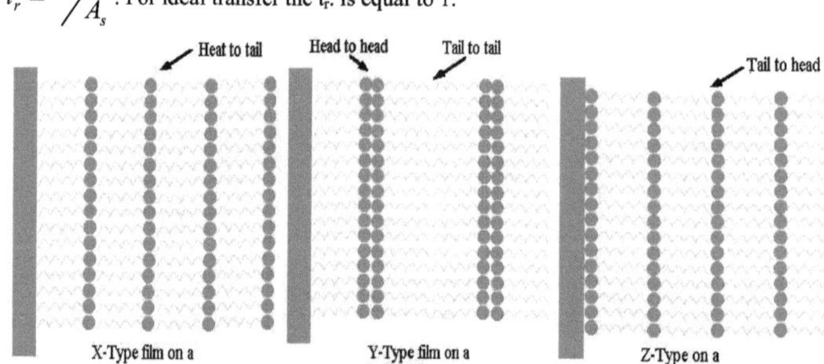

Fig. 11 Multilayers of X, Y and Z-type Langmuir-Blodgett (LB) films onto solid substrates.

For an ideal Y-type Langmuir-Blodgett multilayer system, the transfer ratio for both upstroke and downstroke are constant and equals to unity. However for an ideal X-type Langmuir-Blodgett multilayer system, the transfer ratio is always one for the downstroke and zero for the upstroke. Where as for an ideal Z-type Langmuir-Blodgett multilayer system, the transfer ratio is always zero for the downstroke and one for the upstroke. However deviation from the ideal behaviour is generally observed. Despite intense research in this field the mechanisms of Langmuir-Blodgett (LB) films are yet to be understood clearly and it is certain that the last word has not yet been spoken

4. Characterization of Langmuir-Blodgett films:

The quality of the Langmuir-Blodgett films largely depends on the parameters like pH of subphase, temperature, speed of dipping, way of spreading the materials, surface pressure of lifting, speed of compression of the barrier. The characteristics of LB films are studied by UV-Vis absorption spectroscopy, Flourescence spectroscopy, X-ray diffraction, Electron and neutron diffraction, Atomic force microscopy (AFM), Scanning electron microscopy (SEM), Electron Spin Resonance (ESR), Raman Spectroscopy, Optical harmonic generation, Infra Red Spectroscopy, Pyroelectric, AES, SIMS, ESCA, Surface potential, Scanning transmission microscopy (STM), Brewster angle microscopy (BAM), Transmission electron microscopy (TEM) etc.[1]

5. Recent research on Langmuir-Blodgett (LB) films:

Langmuir-Blodgett film is prepared by transferring floating organic monolayer (Langmuir monolayer) onto solid substrate. The properties of these layers can be controlled with ease, by changing various LB parameters. Also the properties of these layers are strongly dependent on the individual type of molecule.[34] A combination of innovative chemistry and a carefully engineered instrument (a Langmuir trough) can result in a high quality monomolecular assemblies displaying a high degree of structural order.

Research on Langmuir-Blodgett (LB) films in recent years have been summarized by Gaines,[20] G. G. Roberts,[1] I. R. Peterson,[1] A. Ulman,[1] R. H. Tredgold [1] and more recently by M. C. Petty.[1]

The 'modern' period of LB film research dated from the early 1970s with the work of Kuhn and his group. It was probably no coincidence that this was also the beginning of the microelectronics revolution when the technological potential of ultraminiaturised fabrication techniques in general, and thin-film deposition techniques in particular, was first appreciated. In a series of imaginative experiments, Kuhn and coworkers demonstrated that, under suitable conditions, the LB deposition process was capable of assembling molecules into a stable, well defined functional structure. Firstly, they showed that individual monolayers could be prepared with extremely high uniformity.[35] Secondly, they showed that film molecules deposited on distinct immersion-withdrawal strokes could retain the relative position over long periods of time,[36-38] even when the films are subjected to considerable perturbation by cleavage and reassembly.[36,39] Thirdly, they demonstrated that the process of exciton transfer between two different molecular species could be 'fine-tuned' by adjusting the distance between them to subnanometre precision. This work appears to have been the first concrete research activity towards functional systems on a molecular scale.

However it is only the last decade or so that extensive fundamental and applied research potential of monomolecular assemblies prepared using LB technique has been appreciated. Their precise thickness, coupled with the high degree of control over their molecular architecture has now been firmly established a role for these organic multilayer LB films in thin film technology. The LB technique of film deposition is one of the few thin film technologies that actually permit the manipulation of materials at the molecular level.

There are numerous potential applications of LB films in various areas of surface science; besides their potential use in molecular electronic devices,[5] optoelectronics,[40] optical signal processing,[3] digital optical switching devices [4] and optical recording,[41] etc one may also take advantage of their striking nonlinear properties in the field of nonlinear optics.[6]

The natural orientation features of monolayers and the degree of control over molecular architecture provide good reason for utilizing LB films in applications such as acoustic surface wave devices, infra-red detectors and optoelectronics where materials with interesting non-centrosymmetric structures are required.[42-44] For device fabrication purposes it is often preferred to use a thin film rather than a bulk crystal. In order to avoid the symmetry inherent with Y-type films it would appear that X or Z type deposition profiles are necessary, then a permanent polarization may be produced with a strong component in a direction perpendicular to the substrate. But it is observed that the films deposited in this way with their dipoles aligned in a common direction, are invariably of poorer quality than Y-type layers. However, using the constant perimeter barrier alternate layer trough it is possible to achieve a Y-type film where the contributions of adjacent molecules do not cancel. For example, by depositing an acid and an amine whose dipole moments have opposite sense, alternate layer LB films can be produced which may exhibit the desired non-linear properties. A possible method of improving the structure and degree of order in multilayer assemblies is to use electric or magnetic fields to align molecules.

A new approach to second-order nonlinear optical (NLO) materials is reported recently [45] by Professor F. Aroca, Material and Surface Science research group, University of Windsor, Canada, where chirality and supramolecular organization play the key roles. Langmuir-Blodgett (LB) films of a chiral hellcene are composed of supramolecular arrays of molecules. The chiral supramolecular organization makes the second order NLO susceptibility about 30 times larger for the nonracemic materials than the racemic materials with the same chemical structure. The susceptibility of the nonracemic films is a respectable 50 picometers per volt, even though the helicene structure lacks features commonly associated with high nonlinearity. Susceptibility components that are allowed only by chirality dominate the second-order NLO response[45]. Professor J. Kawamata and his group in the Department of Chemistry, Faculty of Science, Yamaguchi University, Japan demonstrated Nonlinear Optical Property (NLO) of Langmuir–Blodgett films consisting of metal complexes.[46]

It is widely recognized that pyroelectric materials have considerable advantages over narrow bandgap semiconductors, as detectors of infrared radiation. There are several advantages that make LB films particularly attractive candidates for pyroelectric devices. Professor R.C.Kapan and his group, Balikesir Üniversitesi Fen-Edebiyat Fakültesi, Turkey, are doing extensive research on the pyroelectric properties of Langmuir-Blodgett (LB) films incorporating ions and other materials.[47] They demonstrated the pyroelectric properties of LB films in a series of works. The most important of these is that the sequential deposition of single monolayers enables the symmetry of the film to be precisely defined; in particular, layers of different materials can be built-up to produce a highly polar structure. Secondly, the polarization of an LB film is 'frozen-in' during deposition and it is therefore not necessary to subject the film to a poling process. The third advantage is that the LB technique uses amphiphilic organic materials which possess low permitivities (ε) and the figure of merit for voltage responsively p/ε (where p is the pyroelectric coefficient) is expressed to be large. Finally the LB method enables the preparation of much thinner films than usually attained by normal conventional techniques. The pyroelectric coefficients for multilayer acid / amine LB films can be about 10 $\mu Cm^2 K^{-1}$ and depend on the thermal expansion coefficient of the substrate, indicating that there is a significant secondary contribution to the measured pyroelectric response.[48-50]

LB films can be used in conjunction with more conventional microfabrication techniques to extend the capabilities of micro-electronic devices, as constituents of chemical sensors and transducers, and for optical applications. In some of these proposals the film is merely used as a spacer layer of extremely uniform thickness, for example as a gate insulator in a field effect transistor,[51] a microphone [52] or as a hyperfiltration layer.[53] In others it is used as a passive support to anchor active molecules, for example enzymes in biosensors. There are three categories where the film molecules themselves perform an active function, for example chemical, in microlithographic resists;[54,55] electronic, in high-density bulk memories [56] or optical, in switches, modulators and non-linear signal processing.[57]

Another use of LB film is in the field of membranes.[52] The similarity between the LB film structure and naturally occurring biological membranes suggests that the former may be exploited as selective barriers for a variety of molecular or ionic species. Recent research has important implication for other application areas such as separation, reverse osmosis and dialysis.[53]

Langmuir-Blodgett (LB) film deposition of metallic nanoparticles and their application to electronic memory structures have been reported by Petty et.al.[58]

The possibility to transfer LB films of perfluorinated ionomer polymers onto solid support for different electrochemical applications were demonstrated by Professor M. K. Ram, Fractal Systems Inc, USA.[59-62] Among the perfluorinated ionomers, Naflon (cation exchanger) has received a lot of attention in recent years owing to its wide range of applications in fuel cells, chemically modified electrodes, sensors etc.[63-72] LB films of perfluorinated ionomers have evidenced interesting electrochemical properties such as a lower charge transport phenomenon than the recasted films even if the amplification of the signal is still good compared to the recasted films.[60]

An interesting application about the possibility to incorporate biological molecules such as cytochrome-C has been recently demonstrated.[62] In this particular case the isotherm of pure cytochrome-C is characterized by a rather broad trend showing an increase in the π-A plot with a collapse pressure of about 16 mN/m. This trend and the relatively low collapse pressure are typical of Langmuir isotherms of proteins and appears similar, for instance, to those recently reported in literature for cytochrome p450 and other enzymes.[73,74]

An interesting work about the formation of LB films of DNA has been recently reported by Professor X. Hou, Key Laboratory for Supramolecular Structure and Materials of Ministry of Education, China.[75] They fabricated LB films of a complex formed between didodecyldimethylammonium and DNA (DNA-DDDA) and investigated the interactions of this comlex with three dyes such as acridine orange, ethidium bromide and prophyrine derivatives (TMP$_y$P). Another application about the utilization of LB technique for the fabrication of nano-organized systems for detection of NADH has been reported by Mecheri et. al.[76] Gold modified electrodes of phospholipids layer containing tetramethylbenzidine (TMB) for NADH has been prepared. Phospholipids are widely used in biosensors technology to mimic biological membrane.[77]

The biomineral calcium oxalate monohydrate (COM) is a major component of urinary stones. An understanding of the processes that lead to COM precipitation in the urinary tract can lead to the prevention or treatment of urinary stone disease. In collaboration with Professor Saeed Khan in the Department of Pathology and Professor Laurie Gower in Materials Science and Engineering, Prof. D. R. Talham, Department of Chemistry, University of Florida, USA, are studying Langmuir-Blodgett (LB) films of phospholipids to model domains in biological membranes for studies of COM precipitation and adhesion. They are interested in studying how the chemical and organizational properties of the model membrane affect the formation of COM crystals at the membrane interface.[78-80]

Professor M. C. Petty and his group in the School of Engineering, Centre for Molecular and Nanoscale Electronics, Durham University, is involved in the study of thin films of organic molecular compounds produced by a variety of techniques such as spinning, thermal evaporation, electro-deposition, self-assembly and Langmuir-Blodgett technique. They are concentrating mainly on the fabrication of electronic devices e.g., solar cells, electroluminescent displays and field effect transistors using these films. Work is going on the investigations of chemical (gas, vapour and liquid) sensors, self-organised structures and nanoscale devices.[81-83] The group includes colleagues within Durham, other academic establishments in the UK and abroad and throughout industry.

Sensor is a device, which provides direct information about the chemical composition of its environment.[84,85] It consists of a physical transducer and a selective sensing layer. It is invariably provided by a material in which some selective interaction of the species of interest takes place that results in the change of some physical parameters such as electrical current or potential or conductivity, intensity of light, mass, temperature etc.

The combination of synthetic chemistry with the molecular engineering capability of the LB technique makes organic multilayered system interesting candidates for sensors.[86] There are numerous physical properties on which LB sensing system can be based. Examples include resistivity changes, electro chemical phenomena, optical effects etc. However, main challenges are the development of new sensors and in the production of cheap, reproducible and reliable devices with adequate sensitivities and selctivities.

Many different types of organic materials have been used for gas sensing. These include porphyrins, phthalocyanines and insulating and conducting polymers.[87,88]

Electrochemical sensors (i.e. chemiresistors) have been widely investigated, also studies of mass and optical changes have been undertaken to gain a better understanding of the polymer / gas interactions. Non-conductive polymers (i.e. those not containing π -bonds) have been used as mass sensors, thermal sensors, optical sensors and dielectric sensors.[89,90]

Polypyrrole (PPy) was widely used as various kinds of sensors depending on their transducing mechanism, including mass sensor,[91] potentiometric sensor,[92-94] potentiometric humidity sensor,[95] amperometric sensor,[96] amperometric biosensor [97-99] and conductometric sensor.[100-103]

Professor Sergey N. Shtykov and his research group, Saratov State University, Russia, showed a novel application of Langmuir–Blodgett films as modifiers of piezoresonance sensors [104]

Professor B. D. Malhotra, Biomolecular Electronics & Conducting Polymer Research Group, National Physical Laboratory, India and others demonstrated that PPy films deposited by LB technique may be used selectively to detect ammonia.[105-107] Professor Milella and Penza, Materials and New Technologies Unit, Italy,[108-111] carried out their systematic studies on PPy based surface acoustic wave (SAW) sensors. LB films of PPy was also tested in some organic vapours [112,113] and found that methanol and ethanol vapour causes an increase in the resistance of the film, which was rapidly reserved in air at room temperature. Langmuir and Langmuir-Blodgett (LB) films of some semi-amphiphilic alternating N-hexylpyrrole-thiophene AB copolymer (PHPT) were investigated as a promising material for sensing.[114] Milella and Penza also demonstrated Carbon nanotubes-coated multi-transducing sensors for VOCs detection.[110]

Artificial tongue can also be made from the sensor array of conducting polymer Ultrathin LB films. Professor A. R. Riul, Universidade Estadual Paulista, Brazil,[115] made an artificial tongue composed of four sensors made from Ultrathin films deposited onto gold interdigitated electrode by LB technique. This tongue was able to detect four basic tastes (salty, sour, sweet and bitter), in addition to detecting inorganic contaminants in ultrapure water and identifying different brands of coconut water. Professor Osvaldo N. Oliveira, Jr and his group in the Universidade Estadual Paulista, Brazil, are now doing extensive research on Langmuir-Blodgett (LB) films of biologically important materials, sensors and memory devices.[116] They demonstrated wine classification system by taste sensors made from ultra-thin Langmuir-Blodgett (LB) films and using neural networks.

Professor De Souza, Departament de Química, Universidade Católica de Pernambuco, Brazil,[117] made fast response sensors of volatile compounds using PPy thin films in-situ self assembly with a variety of dopants. Grader et. al.[118] modeled a polymer gas sensor in terms of homogeneous diffusion coupled to simple adsorption within a bounded layer.

Bio sensors are devises capable of retrieving analytical information from the operational environment by utilizing biological components as part of the sensor. Biosensors use biological molecules, mainly enzyme, lipids etc as the recognition elements.

Professor Hou, Central China Normal University, China,[119] demonstrated a novel biosensor using LB technique to the mixtures of odorants in various environmental conditions.

A novel optical nanosensor using a support bilayer lipid membrane (SBLM) has been recently proposed by Professor O. Worsfold, Frontier Research Division, Fujirebio Inc., Japan.[120] In this work LB and Layer-By-Layer (LBL) techniques have been combined to obtain highly ordered nanostructure. This work is particularly significant due to the importance of sensors for biological agents in vivo and/or in vitro.

A heptamer linear RGD (acridine-glysine-asparate) containing peptide was covalently attached to a BODIPY (2-(4,4-difluoro-5, 7-diphenyl-4-bora-3a, 4a-diaza-s-dodecanoyl)-1-hexadecanoyl-glycero-3-phospho ethanolamine, donor) lipid dye and utilized as an optical biosensor. A second BODIPY (4,4-difluoro-5-(2-thienyl)-4-bora-3a,4a-diaza-s-indacene-3-dodec-anoic acid, acceptor) lipid dye was integrated into the SBLM, enabling the signal amplification via a Forster resonance energy transfer (FRET) mechanism. The result indicates the possibility to detect HUVEC at a concentration of 1000 cells ml^{-1}. The sensitivity obtained by this method is similar to polymerized chain reaction (PCR) technique methods but less sensitive than flow cytometric techniques.[121,122]

A glucose sensor consisting of a conductive polypyrrole membrane and a lipid LB film has been investigated.[123] Professor M. Rikukawa, Department of Chemistry, Sophia University, Japan,[124] fabricated a similar biosensor comprised of a lipid-modified glucose oxidase and conducting PPy LB film for detecting glucose. Professor T. N. Misra, IACS, India,[125] prepared a stable monolayer of alcohol dehydrogenase (ADH) enzyme by spreading an aqueous solution of ADH on a water subphase containing stearic acid monolayer. This ADH-stearic acid monolayer has been successfully transferred onto a conducting polypyrrole –coated glass electrode by the LB technique. This LB-immobilized polypyrrole-mediated enzyme electrode can be used as an ethanol sensor. Novel enzyme based micrelectrochemical devices, based on changes in the conductivity of polypyrrole layers, were developed for biosensing of NADH and penicillin.[126]

Biotinylated protein, such as DNA and the photoactive protein, phycoerythrin, are stably attached to PPy monolayer film via a bridging streptavidin protein.[127,128] The strategy was to biotinylate one dimensional elecro-active polymers and use bridging streptavidin protein on LB organized films. Polycation conducting polymers, oxidized polypyrrole possess the ability to form complexes with polyanionic DNA molecules largely through electrostatic interactions. These studies present a way to detect DNA hybridization using polypyrrole DNA chip. The immobilization of long DNA on a LB film with the help of Zn coordination has been demonstrated.[129]

Since the discovery of semiconducting behaviour in organic materials, there has been considerable effort aimed at exploiting these properties in electronic and optoelectronic devices. The simple fabrication techniques for polymers have attracted several companies to work on polymer transistor application such as data storage and thin film device arrays to address liquid crystal displays.[130]

Semiconducting LB films have also been used in conjunction with the inorganic semiconductors (eg. Si, GaAs) to form metal / semiconductor / metal structures. Perhaps the simplest example is that of a diode. Here, the LB film is sandwiched between metals of different work functions. In the ideal case, an n-type semiconductor should make an ohmic contact to a low work function metal and a rectifying Schottky barrier to a high work function metal.[131,132] The reverse will be true for a p-type semiconductor. Phthalocyanine LB film is a p-type semiconductor and the aluminium / LB film interface provides the rectifying junction.[132]

There is much current interest in the possibility of observing molecular rectification using monolayer and multilayer. This follows the prediction of Aviram and Ratner that an asymmetric molecule containing a donor and an acceptor group separated by a short σ -bonded bridge (allowing tunneling) should exhibit diode characteristics.[133] There have been many attempts to obtain this effect in LB films.[1] However, the demonstration that the rectification is affected by a change in the film structure (by bleaching)[134] and a recent report of rectifying behaviour with symmetrical gold electrodes[135] suggest that true molecular rectification is achievable.

Experimental and theoretical confirmation of anion-induced dipole reversal in cationic dyes due to molecular rectification have been demonstrated by Professor G. J. Ashwell, and S. H. B. Kelly, The Nanomaterials Group, Centre for Photonics & Optical Engineering, Cranfield University, UK.[136]

Professor Robert M. Metzger in the Department of Chemistry, University of Alabama, USA reviewed the application of Langmuir-Blodgett (LB) films as unimolecular rectifier.[136]

Another interesting application of the LB technique regarding the fabrication of thin films of copper phthalocyanine derivatves as field effect transistor (FET). It is well known that pthalocyanine derivatives are very promising organic semiconductor materials due to their chemical and thermal stability. Among phthalocyanine, copper phthalocyanine derivatives have been utilized as organic field effect transistors (OFETs). The FET performances of the LB films of phthalocyanine has been tested by I-V curves acquired from devices operating in accumulation mode. It has been observed that to improve the carrier mobility of PC films, the arrangement through a more highly ordered film with improved interaction distance and $\pi - \pi$ interaction and decreases the effects to result in better quality of the LB films throughout the FET channel.

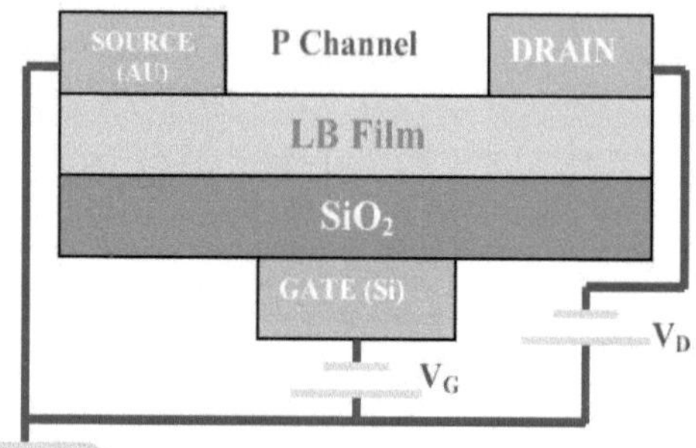

Fig. 12 Schematic of OFET of pthalocyanine LB films

Recently an interesting work about a physico-chemical investigation of carboxylic ionophores and phospholipids for application as ion selective field effect transistor (ISFET) has been reported.[137]

Cui et. al.[138] demonstrated depletion-mode n- channel organic field effect transistors (OFETs) based on naphthalene-tetracarboxylic-dianhydride (NTCDA) transistor, n-type NTCDA acts as active channel material due to its high mobility of 0.06 $cm^2V^{-1}S^{-1}$, and p-type conducting polymer Polypyrrole performs as the source and drain. Koezuka et. al.[139] fabricated a FET by using two different kinds of conducting polymers, Polypyrrole and polythiophene.

Fabrication of Polymer Langmuir-Blodgett Films Containing Regioregular poly(3-hexylthiophene) for application in Field-Effect Transistor (FET) have been reported by Matsui et. al.[140]

Su et. al demonstrated Thin-Film Transistors based on Langmuir-Blodgett Films of Heteroleptic Bis(phthalocyanine) Rare Earth Complexes.[141]

Organic field-effect transistors based on Langmuir-Blodgett films of an extended porphyrin analogue – Cyclo[6] pyrrole and neutral long-chain TCNQ derivatives have also been reported.[142,143]

Organic thin films have also been used as the semiconducting layers in FET devices.[144] A significant increase in the carrier mobility has been reported over the last fifteen years using LB films of organic materials. [145-147]

Materials	Mobility $(cm^2V^{-1}s^{-1})$
Polythiphene	10^{-5}
Polyacetylene	10^{-4}
Phthalocyanine	$10^{-4} - 10^{-2}$
Thiophene oligomers	$10^{-4} - 10^{-1}$
Pentacene	$10^{-3} - 3$
C_{60}	0.3
Organometallic dmit complex	10^{-1}

Table 1: Field effect carrier mobilities of FET based on organic semiconductors [145-147]

Thin-Film Transistors Based on Langmuir-Blodgett Films of Heteroleptic Bis(phthalocyanine) Rare Earth Complexes have also been reported.[148]

Solution processed Langmuir-Schäfer and cast thin films of poly (2,5-dioctyloxy-1,4- phenylene-alt-2,5-thienylene) were also investigated as transistor active layers.[149] The study of their field-effect properties evidences that no transistor behavior can be seen with a cast film channel material. This was not surprising considering the twisted conformation of the polymer backbone predicted by various theoretical studies. Strikingly, the Langmuir-Schäfer (LS) thin films exhibit a field-effect mobility of 5×10^{-4}

$cm^2/V.s$, the highest attained so far with an alkoxy-substituted conjugated polymer. Extensive optical, morphological, and structural thin film characterizations support the attribution of the effect to the longer conjugation length achieved in the Langmuir-Schäfer (LS) deposited film, likely due to an improved backbone planarity. This study shows that a technologically appealing deposition procedure, such as the LS one, can be exploited to significantly improve the inherently poor field-effect properties of twisted conjugated backbones. This achievement could promote the exploitation for electronic, and possibly sensing, applications of the wealth of opportunities offered by the alkoxy substitution on the phenylene units for convenient tailoring of the phenylene-thienylene backbone with molecules of chemical and biological interest.

Polymeric light emitting diodes (PLEDs) have shown much interest worldwide since the discovery of electroluminescence (EL) in a thin poly(phenylene vinylene) layer by Friend and coworkers in 1990.[150,151] Conducting polymers have been thought of as potential luminescent materials for replacing inorganic light emitting materials when used in large area, light weight, flexible displays. The main advantage of these materials over conventional luminescent materials are the tuning of wavelength emitted by chemical modification, low operating voltage, flexibility, easy processing, low cost, possibility of making large device and output colours in whole visible spectrum. Several p-doped conducting polymers have been tested in LB films [152-156] and have been used as hole injecting electrodes, like polypyrrole, polythiophene derivatives and polyaniline, which have high work functions, providing low barriers for hole injections.

The relatively ordered nature of LB arrays of molecules may be exploited in light emitting structures. For example, the preferential alignment of the molecules can result in polarized light emission.[157,158] Langmuir-Blodgett (LB) film deposition can also be used to control the positions of the luminescent species within metal mirror microcavities.[159]

The technique of using insulating LB layers to improve the efficiency of the OLEDs seems generally applicable and it has also been used to enhance the operation N, N' -diphenyl-N, N' -bis (3-methylphenyl) 1,$1'$ -biphenyl-4, $4'$ -diamine (TPD)/tris-(8-hydrxyquinoline) aluminium (Alq_3) bilayer devices in which both the emissive layer (Alq_3) and the hole transport layer were deposited by thermal evaporation. To improve the lifetime of OLEDs several experiment on MEH-PPV LB films [160] have been done. More recent work has been focused on hybrid of pyridine and oxadiazole which lead to a further increase in the electron-injection.[160]

Many conducting polymers such as polyacetylene, polythiophene, polyindole, polypyrrole, polyaniline etc have been reported as electrode materials for rechargeable batteries.[161] Recently Langmuir-Blodgett (LB) technique was used to deposit polypyrrole thin films.

Photoconductivity involves enhancement of the electrical conductivity of the material by the absorption of a suitable photon. It finds wide range of applications in electronics, for example auto brightness control (ABC) circuits in TV sets, camera shutter, car dimmers, street light control, auto gain control in transceivers electrophotography etc. Electrochemically produced polypyrrole Langmuir-Blodgett (LB) films (band gap 3.2 eV) after sensitization can be anticipated to exhibit good photoconductivity.[162]

Liu et. al.[163] fabricated a polymer based capacitor, using polypyrrole and poly (3,4-ethylenedioxythiphene) poly(styrenesulfonate) as a semiconductor and gate layer. Dielectric polymer, polyvenylphenol, was applied as the insulator to the device. Composite electrodes for supercapacitors were prepared via chemical polymerization of pyrrole on the surface of a porous graphite fiber matrix.[164]

Ferreira et. al.[165] obtained homogeneous and adherent polypyrrole films when the steel surface was treated with dilute nitric acid. It was shown that the reaction of nitric acid with the iron led to the formation of iron nitride (Fe=N) species, which were responsible for the formation of homogeneous and adherent polypyrrole films on iron.

Professor Jager, Biomolecular and Organic Electronics, IFM, Linköping University, S-581 83, Linköping, Sweden,[166,167] made the first attempt towards all –polymer electrochemical microacuators that had all the electrodes necessary for the acuation – the working, counter, and reference electrodes-on-chip.

Rectifying junctions are the basic elements of many electronic components. Rectifying junctions such as polymer p-n junction and schottky junction based on

Langmuir-Blodgett (LB) films have been studied widely.[168] Aizawa et. al.[169] have demonstrated that anion doped polypyrrole / polythiphene can behave as n-type semiconductor. Metal-insulator-semiconductor (MIS) structures were fabricated by vacuum deposition of various metals like indium, aluminium and tin on LB films of cadmium stearate ($CdSt_2$) obtained on polypyrrole films deposited on indium-tin-oxide glass.[170] The value of the dielectric constant of the insulating $CdSt_2$ LB film has been found to be 1.84 which is in good agreement with the value reported earlier. Organic heterostructures have been fabricated by alternating deposition of mono- and multilayers of undoped poly (3-hexylthiophene) and doped polypyrrole prepared by Langmuir-Blodgett (LB) technique.[171] Electrical rectifying devices (Zener diode) have been prepared using a two layer configuration, consisting of a p^+-doped semiconducting polymer Polypyrrole or poly (3-methylthiophene) layer and a n-type multilayer structure of CdSe and 1,6-hexenedithiol.[172]

In recent years efforts are on to fabricate crystalline materials of nano order dimension using LB technique. Nanocrystalline materials of various metallic compounds have been fabricated using LB technique. FeS, ZnS, PbS, PbSe and several other metallic compounds in the nanocrystalline order have been formed in the LB films using various methods.[173] Various light sensitive materials form nanocrystalline aggregates in the LB films under photo induced self organization.[174] Ageing effects on the morphology and structure of LB films have also been studied and it is observed that in some cases the organized molecular assemblies in LB films tend to form small crystalline domain having the dimension of micrometer to nanometer order.[175]

A great deal of research are currently being done on various technical applications of nanoscale particles organized in LB films and have found their potential applications in optical,[176] electronic [177] and biosensing devices.[120] The LB films of alkane thiol-capped gold nanoparticles, formed under high surface pressure have found their potential applications in molecular rectifier devices.[178] CdSe nanocrystal based semiconductor-insulator-metal tunnel diodes have been developed.[179] Nanometer scale metal-LB films-metal structure is realized with an Atomic Force Microscope combined with STM, in which increase in conductance at any point in LB films can be induced by application of a voltage pulse.[180]

Professor M. Sastry and his group of NCL, India, demonstrated nanoscale assembly using hydrophobic interactions to organize nanoclusters with controllable spacing [181] as well as on high aspect ratio structures like carbon nanotubes.

A convenient method to produce highly uniform ZnS Supercrystals of Uniform Nanorods and Nanowires, and the Nanorod-to-Nanowire oriented transition by Langmuir-Blodgett (LB) technique have been reported by Narayan Pradhan and Shlomo Efrima [182] of Ben-Gurion University, Israel.

Two-Dimensional Pressure-Driven Nanorod-to-Nanowire Reactions in Langmuir Monolayer at Room Temperature was demonstrated by Acharya and Efrima.[183]

The LB multilayers have been used to grow chalcogenide semiconducting nanoclusters in organic fatty acid matrix through post deposition treatment.[184,185] The interest in this approach is primarily because of the high degree of molecular order present in LB multilayers, which is expected to assist in achieving better control over the geometry, shape, size and distribution of the nanoclusters. This approach has recently been extended to develop CdS nanoclusters containing polyaniline LB films using precursor polyaniline-cadmium arachidate (PANI-CdA) composite LB multilayer.[186] The composite monolayer methodology in which a nonamphiphilic polymer is mixed with suitable metal ion containing amphiphilic molecule provides a simple approach to develop these polymer-based precursors.[187,188] Recently, the Langmuir monolayer behaviour of poly (*o*-anisidine) polyanilines carbon nano tubes, Langmuir-Blodgett films were prepared, characterized and used as gas sensor by Ram et. al.[189-192]

Carbon nanotubes (CNTs) show a great promise for engineering applications due to their high streangth and stiffness, as well as unique physical and electrical properties. CNTs find applications in scanning probes,[193-195] electron field emission sources, acquators, nano-electronic devices, medical devices, hydrogen storage and nanocomposites materials.[196-199] More recently, nanocomposites of multiwalled carbon nanotubes (MWNTs) embedded in poly(*o*-anisidine) (POAS) and poly (*o*-toludine)-

MWNTs were synthesized and Langmuir-Blodgett (LB) films were fabricated.[200] Professor Yoshiaki Imaizumi, Department of Micro-Nano Systems Engineering, Nagoya University, Japan, demonstrated the preparation of self-assembled giant carbon nanotube construction using Langmuir–Blodgett films.[201]

Substitution controlled molecular orientation and nanostructure in the Langmuir–Blodgett films of a series of amphiphilic naphthylidene-containing Schiff Base derivatives have been reported by Professor Tifeng Jiao, CAS Key Laboratory of Colloid and Interface Science, China.[202]

An easy and reliable method for the production of patterned monolayers of Cobalt nanoparticles using Langmuir-Blodgett (LB) technique was reported by Kim et. al.[203] Where Transmission Electron Microscopy (TEM) was used to show that with increasing surface pressure, the Cobalt nanoparticles become well-organized into a Langmuir monolayer with a hexagonal close-packed structure. By controlling the pH of the subphase, it was found that a monolayer of Cobalt nanoparticles with long-range order could be obtained. Further, by transferring the Langmuir monolayer onto a poly-(dimethoxysilane) (PDMS) mold, the selective micropatterning of the cobalt nanoparticles could be achieved on a patterned electronic circuit. The electronic transport properties of the cobalt nanoparticles showed the ohmic I-V curve.

Hongjie Dai et. al. synthesized gram quantities of uniform Ge nanowires (GeNWs) and showed that they could be readily assembled into close-packed Langmuir-Blodgett films potentially useful for future high performance electronic devices.[204]

Silver sulfide (Ag_2S) nanocrystals were prepared in ambient conditions by exposing crystalline polydiacetylene (PDA) Langmuir films at the air-$AgNO_3$ solution interface to H_2S gas.[205]

Two-dimensional arrays of the Platinum nanocrystals were assembled by using the Langmuir-Blodgett (LB) method. The particles were evenly distributed on the entire substrate, and their surface coverage and density could be precisely controlled by tuning the surface pressure. The resulting Platinum LB layers were potential candidates for 2-D model catalysts as a result of their high surface area and the structural uniformity of the metal nanocrystals.[206]

An attractive contribution to the interparticle potential, such as the dipolar potential, plays a significant role in the spontaneous organization of CdSe nanoparticles in the Langmuir-Blodgett (LB) films.[207]

Densely packed exfoliated nanosheet films such as Ti0.91O2, Ti0.8M0.2O2 (M) Co, Ni), Ti0.6Fe0.4O2, and Ca2Nb3O10 on solid substrates were prepared by the LB transfer method without any amphiphilic additives at the air-water interface.[208] Nanosheet crystallites covered nearly 95% on the solid surface with minimum overlapping of nanosheets. The LB transfer method of the Ti0.91O2 nanosheet monolayer film is applicable for not only hydrophilic substrates such as quartz, silicon, indium-tin oxide (ITO), and glass but also the hydrophobic Au surface. On the basis of these points, the LB transfer method has advantages compared to the alternating layer-by-layer method, which makes use of oppositely charged polyelectrolytes such as poly(ethylenimine) (PEI).

Professor I. R. Peterson and his group in the Coventry University, UK, showed that Langmuir-Schaefer (LS) monolayer films of fullerene-bis-[4-diphenylamino- 4″ -(N-ethyl-N- 2‴ -ethyl)amino- 1,4-diphenyl-1,3-butadiene] malonate, **1**, sandwiched between two Au electrodes, exhibit pronounced current asymmetries (rectification) between positive and negative bias at room temperature, with no decay of the rectification after several cycles. The device shows symmetrical through-space tunneling for a bias up to 3 V, and asymmetrical, unimolecular, "U" type rectifier behavior in the voltage range from 3.0 to 5.4 V, with rectification ratios up to 16.5. The rectification is ascribed to the asymmetric placement of the relevant molecular orbitals, with respect to the metallic electrodes.[209] The rectification ratios are reasonable and persist over many cycles of observation.

Langmuir-Blodgett Films of Poly [2-methoxy-5-(*n*-hexyloxy)-*p*-phenylenevinylene] showed reversible photoconducting properties.[210] The efficiency of poly (acrylic acid) and poly(methacrylic acid) in gluing monolayers of calix[6]arene **1**, and in controlling the barrier properties of resulting LB films, has been found to be very sensitive to pH. Low pH favors the formation of highly viscous monolayer, a high density polymer chains associated with the glued bilayer, relatively high barrier properties, and relatively high permeation selectivity. At high pH, fewer polymer chains become incorporated into the glued bilayer, and the barrier properties and permeation selectivity are relatively low. A model that is consistent with these features, which is based on a combination of ionic and hydrophobic interactions, has been proposed.[211]

Secondary Structure of Organophosphorus Hydrolase in Solution and in Langmuir-Blodgett Film was studied by Circular Dichroism Spectroscopy.[212] It was observed that there was a decrease in the percentage of the R-helical and an increase of ,-strands with the change of pH or temperature.

Anomalous Phase Behavior in Langmuir Monolayer of Monomyristoyl-*rac*-Glycerol at the Air-Water Interface has been reported by Md. Nazrul Islam and Teiji Kato, Department of Applied Chemistry, Faculty of Engineering, Utsunomiya University, Japan.[213]

The photoinduced electron transfer in several structurally different phytochlorin-fullerene and porphyrin-fullerene dyads has been studied in polar and non polar solvents using femtosecond fluorescence up-conversion and pump-probe transient absorption techniques. This has been a continuation of the long-term fundamental research of the artificial molecular systems capable of performing photoinduced electron transfer. Professor N. V. Tkachenko and his group, Institute of Materials Chemistry, Tampere University of Technology, Tampere, Finland, demonstrated vectorial photoinduced electron transfer in alternating Lanbmuir-Blodgett films of phytochlorin-[60]fullerene dyad and regioregular poly(3-hexylthiophene).[214,215] The films were studied using time-resolved Maxwell Displacement Charge measurements and demonstrated their functionality in different sample arrangements.

Single Molecule Detection (SMD) using Surface-Enhanced Resonance Raman scattering and Langmuir-Blodgett monolayers have been demonstrated by Professor R. F. Aroca and O. N. Oliveira Jr.[216-218] In this work the Langmuir-Blodgett (LB) technique has been used to obtain the Surface-Enhanced Resonance Raman scattering (SERRS) spectra of single dye molecules in the matrix of a fatty acid. Mixed monolayers of the dye material have been fabricated on silver islands substrates with a concentration of the probe molecule of one molecule per micron square.

6. Conclusions:

In the light of the above reviews it is obvious that the Langmuir-Blodgett (LB) technique is a way of making ultrathin organic layers with a combination of characteristics not found in any other method. This technique allows intriguing molecular architecture to be built-up on solid surfaces. This method also allows elegant experiments to be undertaken in the research laboratory that can provide valuable insight into the physical processes that underpin the device operation. This work will also pave the way for the development of molecular scale electronic (and ionic) devices, which emulate natural processes. Development of LB films for practical applications is a challenge, requiring an interdisciplinary outlook which neither balks at the physics involved in understanding assemblies of partially disordered and highly anisotropic molecules nor at the cookery involved in making them. LB films have a unique potential for controlling the structure of organized matter on the ultimate scale of miniaturization and must surely find a niche where this potential is fulfilled. Therefore it is highly appropriate to make a great stride in these important and promising areas of research, which can provide a conceptual understanding with wide opportunity of technological applications.

7. Acknowledgement:

The author is grateful to UGC and CSIR, Govt. of India for providing financial assistance through UGC minor project Ref. No. F.1-1/2000(FD-III)/399 and Sanction Ref. No. F.31-30/2005 337 (SR) and CSIR project Ref. No. 03(1080)/06/EMR-II. We are also grateful to H. Leman for BAM image measurement.

8. References:

1.

(i). G. Roberts, Langmuir-Blodgett Films, Plenum Press, New York, 1990
(ii). I. R. Peterson *J. phys. D: Appl. Phys* 23 379 (1990)

(iii). A. Ulman, An Introduction to Ultrathin Organic Films: From Langmuir-Blodgett Films of Self assemblies, Academic Press, New York, 1991

(iv). R. H. Tredgold, Order in Organic Thin Films, Cambridge University Press, Cambridge, UK, 1994

(v). M. C. Petty, Langmuir-Blodgett Films: An Introduction, Cambridge University Press, Cambridge, UK, 1996

2.

(i) A. R. Jr., H. C. De Souza, R. R. Malmegrim, D. Santos Jr., A.C.P.L.F Carvalho, F. J. Fonseca, O. N. Oliveira Jr., L. H. C. Mattoso, *Sens. Act. B* **98**, 77 (2004)

(ii) C. E. Borato, A. R. Jr., M. Ferreira, O. N. Oliveira Jr., L. H. C. Mattoso, *Instrum. Sci. and Technol.* **32** 21 (2004)

(iii) M. Ferreira, C. J. L. Constantino, A. R. Jr., K. Wohnrath, R. F. Aroca, J. A. Glacometti, O. N. Oliveira Jr., L. H. C. Mattoso, *Polymer* **44** 4205 (2003).

3.

(i) D. T. Balogh, C. R. Mendoncaa, A. Dhanabalan, S. Major, S. S. Talwar, S.C. Zilio, O.N. Oliveira Jr.; *Material Chem. Phys.* **80** (2) 541 (2003)

(ii) C. R. Mendoncaa, D.S. dos Santos Jr. D.T. Balogh, G. A. Giaconetti, S. C. Zilio, O.N. Oliveira Jr *Polymer* **42** (15) 6539 (2001)

(iii) L.M. Bilinov, S.P. Palto, S. G. Yudin *Appl. Phys. Letts* **80** (1) 16 (2002

4.

(i) Y. Zong, K. Tawa, B. Menges, J. Ruhe, W. Knoll *Langmuir* **21** (15) 7036 (2005)

(ii) L.M. Bilinov, V.M. Fridkin, S.P. Palto, A.V. Sorokin and S. G. Yudin; Thin Solid Films; 284-285 1996 p.474-476

(iii) A. Bune, S. Ducharme L.M. Bilinov, V.M. Fridkin, S.P. Palto; N. Petukhova S. G. Yudin Appl. Phys. Letts.; 67 (26) 1995, 3975-3977

5.

(i). F. de Moura, M. Trisic; *J. Phys. Chem B.* **109** (9) 4032 (2005)

(ii) Z. Biana, K. Wanga, L. Jina, C. Huangh, *Colloids and Surfaces A* **257-258** 67 (2005)

(iii) K. Han, Q. Wang, G. Tang, H. Li, X. Sheng *Thin Solid Films* **476 (1)** 152 (2005)

(iv) K. M. Mayyaa, N. Jainb, A. Goleb, D. Langevin, M. Sastry *J. Coll. Int. Sc.* **270** (1) 133 (2004)

(v) D. Natalia, M. sanpietroa, L. Francoa, A. Bolognesic and C. Bottac *Thin Solid Films* **472** (1-2) 238 (2005)

(vi) G.G. Roberts, M. C. Petty, S. Baker, M.T. Fowlers and N.J. Thomas; *Thin Solid Films* **132** (1-4) 113 (1985)

(vii) T.J. Rece, S. Ducharme, A.V. Sorokin, M. Poulsen *Appl. Phys. Letts.* **82 (1)** 142 (2003)

6.

(i) N. Tancrez, C. Feuvrie, I. Ledoux, J. Zyss, L. Toupet, H. Le Bozec, D. Maury *J. Am. Chem. Soc.* **127**(39) 13474 (2005)

(ii). F. Aroca, T. Verbiest, *Clays Phys. Rev. A* **71** (5) (2005)

(iii). T. Verbiest, S. Sioncke, G. Koeckelberghs *Chem. Phys. Letts.* **404** (1-3) 112 (2005)

(iv) J. W. Boldwin, R.R. Amaresh, I.R. Peterson, J.W.Shumate, P.M. Cava, A.M. Amiri, R. Hamilton, G. Ashwell, R.M. Metzger; *J. Phys. Chem. B* **106(47)** 12158 (2002)

7. D. tabor *J. Colloid. Interface Sci.* **75** 240 (1980)

8. T. Terada, R. Yamamoto, T.Watanabe *Sci. Papers, Inst. Phys. Chem. Research (Tokyo)* **23** 173 (1984)

9. B. Franklin *Phil. Trans. R. Soc.* **64** 445 (1774)

10. J. C.Scott *History Technol.* **3** 163 (1978)

11. J. Shields British Patent No. 3490 (1879)

12. C. F. Gordon-Cumming From the Hebrides to the Himalayas (London: Sampsonhow, Marston, Searl and Rivington) (1876) p. 347.

13. L. Rayleigh *Proc. Soc.* **47** 364 (1890)

14. A. Pockels *Nature* **43** 437 (1891)

15. L. Rayleigh *Phil. Mag.* **48** 321 (1899)

16. I. Langmuir *J. Am. chem. Soc.* **39** 1848 (1917)

17. I. Langmuir *Trans. Faraday Soc.* **15** 62 (1920)

18. K. B. Blodgett *J. Am. chem. Soc.* **57** 1007 (1935)

19. G. L. Gaines *Thin Solid Films* **99** (1983)

20. G. L. Gaines, Insoluble Monolayers at Liquid-Gas Interface (New York: Inter-science) 1966

21. H. Kuhn, D. Mobius, H. Bocner, *Physical Methods of Chemistry*, 1, 577. (1972)

22. B. Maggio, Q. F. Ahkong and J. A. Lucy, *J. Biochem.* **158** 647 (1976)

23. D. O. Shah, J. H. Schulman, *J. Lipid Res.* **8** 215 (1967)

24. G. L. Gains, insolubale Monolayers at the Liquid-Gas Interface, Interscience, New York, 1986

25. J. N. Israelachvili, intermolecular and Surface Forces, Academic Press, 1985

26. B. P. Birks, *Adv. Colloid Interface Sci.* **34** 343 (1991)

27. L. salem, *J. Phys. Chem.* **37** 2100 (1962)
28. J. H. Brooks, A. E. Alexander, radiation of Evaporation by Monolayers, ed. La Mer, V. K. Academic Press, New York, 1962
29. Ter Minassian-Saraga, J. Colloid Sci. **11** 398 (1956)
30. J. H. Brooks, A. E. Alexander, *J. Phys. Chem.* **66** 1851 (1962)
31. D. Vollhardt, T. kato and M. Kawano *J. Phys. Chem.* **100** 414 (1996)
32. M. M. Sacre, E. M. El Mashak, J. F. Tocante *Chem. Phys. Lipids* **20** 305 (1977)
33. A. D. Bangham, W. Br. Mason, *J. Pharmacol.* **66** 259 (1979)
34. Lu Zou, Ji Wang, Violeta J. Beleva, Edgar E. Kooijman, Svetlana V. Primak, Jens Risse, Wolfgang Weissflog, Antal Ja´kli, Elizabeth K. Mann, Langmuir 20 (2004) 2772-2780
35. B. Mann, H. Kuhn *J . Appl. Phys.* **42** 4398 (1971)
36. H. Kuhn *Naturwiss* **54** 429 (1967)
37. D. Mobius *Acc. Chem. Res.* **14** 63 (1981)
38. H. Kuhn *Thin Solid Films* **99** 1 (1983)
39. H. Kuhn, D. Mobius, H. Bucher, *Physical Methods of Chemistry* vol. 1 ed A Weissberger and B Rossiter (New York: Wiley) part 3B chap 7, 1972
40.
   (i) Z. Biana, K. Wanga, L. Jina C. Huangh *Colloids and Surfaces A* **257-258** 67 (2005)
   (ii) K. Han, Q. Wang, G. Tang, H. Li, X. Sheng, *Thin Solid Films* **476** (1) 152 (2005)
   (iii) K. M. Mayyaa, N; Jainb, A. Goleb, D. Langevin, M. Sastry *J.Coll. Int. Sc.* **270** (1) 133 (2004)
   (iv) D. Natalia, M. sanpietroa, L. Francoa, A. Bolognesic and C. Bottac *Thin Solid Films* **472** (1-2) 238 (2005)
41. C. R. Mendinca, D.S. dos Santos Jr., D.T. Balogh, A. Dhanabalan, S. C. Giacometti *Polymer* **42** (15) 6539 (2001)
42. E. Gubbelmans, T. Verbiest, I. Picard; *Polymer* **46** (6) 24 1784 (2005)
43. M. C. Petty Possible applications for Langmuir-Blodgett Films Thin Solid Films 210/211 417 p-1992
44. Zyss J and Chemla D S Nonlinear Optical Properties of Organic Molecules and Crystals vol 1 (Orlando, FL: Academic) p-1987
45.
   (i) F. Aroca, T. Verbiest *K. Clays Phys. Rev. A* **71** (5) 2005
   (ii) T. Verbiest, S. Sioncke, G. Koeckelberghs *Chem. Phys. Letts.* **404** (1-3) 112 (2005)
46.
   (i) T. Higashi, S. Miyazaki, S. Nakamuraa, R. Seike, S. Tani, S. Hayami, J. Kawamata *Colloid Surf. A* **284-285** 161 (2006)
   (ii)J. Kawamata, , , R. Seike, T. Higashi, Y. Inada, J. Sasaki, Y. Ogata, S. Tani, A. Yamagishi *Colloid Surf. A* **284-285** 135 (2006)
47.
   (i) R. Capan, I. Alpa, T. H. Richardson, F. Davis *Materials Letters* **59**(14-15) 1945 (2005)
   (ii) R. Çapan, T.H. Richardson, F. Davis *Mat. Lett.* **59** (14-15) 1945 (2005)
   (iii) R. Çapan, T. H. Richardson, D. Lacey, *Thin Solid Films* **468** (1-2) 262 (2004)
   (iv) R. Çapan, T. H. Richardson, *Mat. Lett.* **58** (25) 3131 (2004)
   (v) R. Çapan, Ba aran, T. H. Richardson, D. Lacey *Mat. Sc. Eng. C* **22**(2) 245 (2002)
   (vi) R. Çapan, T. H. Richardson, D. Lacey, *Thin Solid Films* **415** (1-2) 236 (2002)
   (vii) R. Capan, T. H. Richardson, J. Tsibouklis *Colloids Surf. A* **198-200** 835 (2002)
   (viii) D. Lacey, T. Richardson, F. Davis and R. Capan *Mat. Sc.Eng. C* **8-9** 377 (1999)
48. M. C. petty, in Langmuir-Blodgett Films (G.G. Roberts, ed.), Plenum Press, New York, (1990) p-133
49. D. Lacey, T. H. Richardson, F. davis, R. Caplan, *Mat. Sci. eng. C* **8-9** 377 (1999)
50. T. H. Richardson, in Functional Organic and Polymeric Materials, ed. T. H. Richardson, Wiley, Chichester 181 (2000)
51. K. K. Kan, G. G. Roberts, M. C. Petty *Thin Solid Films* **99** 291 (1983)
52. J. R. Drabble, S. M. Al-Khowaildi *Thin Solid Films* **99** 271 (1983)
53. K. Heckmann, C. Strobl, S. Bauer, *Thin Solid Films* **99** 265 (1983)
54. I. R. Peterson *Proc. IEE* **130** I 252 (1983)
55. B. M. J. Kellner, G. Czornyj *Proc. Symp. On Surface and Colloid Science* (Potsdam, N Y) June 24-28 (Computer Techno) 1985
56. E. G. Wilson *Mol. Cryst. Liq. Cryst.* 121 271 (1985)
57. Hann R A and Bloor D (ed) Organic Materials for Nonlinear Optics Special Publication **69** (London: R. Soc. Chem.) 1989
58. S. Paul, C. Pearson, A. Molly, M. A. Cousins, M. Green, S. Kolliopoulou, p. Dimitrakis, P. Normand, D. tsoukalas, M. C. Petty *Nano Letts.* 3(4) 533 (2003)
59. P. Bertoncello, M. K. Ram, A. Notargiacomo, P. Ugo, C. Nicolini, *Phys. Chem. Chem. Phys.* 4036 (2002)

60. P. Bertoncello, P. Ugo *J. Braz. Chem. Soc.* 517 (2003)

61. P. Ugo, P. Bertoncello, F. Vezza *Electrochim Acta* 3785 (2004)

62. L. M. Moretto, P. Bertoncello, F. Vezza, P. Ugo, *Bioelectrochemistry* 29 (2005)

63. R. S. Yeo *J. Electrochem. Soc.* 533 (1983)

64. B. Tazi, O. Savadogo *Electrochim Acta* 4329 (2000)

65. B. Baradie, J. P. Dodelet, D. Guay *J. electroanal Chem.* 101 (2000)

66. H. S. White, J. Leddy, A. J. Bard *J. Am. Chem. Soc.* 4811 (1982)

67. C. R. Martin, I. Rubinstein, A. J. Bard *J. Am. Chem. Soc.* 4817 (1982)

68. A. D. Buttry, J. M. Saveant, F. C. Anson *J. Phys. Chem.* 3086 (1984)

69. A. D. Buttry, J. M. F. C. Anson J. *A. Chem. Soc.* 4824 (1984)

70. Q. Huang, Z. Lu, J. Rusling, *Langmuir* 5472 (1996)

71. I. Galeska, D. Chattopadhyaya, F. Moussy, F. Papadimitrakopoulos, *Biomacromolecules* 202 (2000)

72. D. G. Fresnadillo, M. D. Marazuela, M. C. Moreno-Bondi, G. Orellana *Langmuir* 6451 (1999)

73. C. Nicolini, V. erokhin, P. ghisellini, C. paternolli, M. K. Ram, V. Sivozhelezov *Langmuir* 3719 (2001)

74. L. Pastorino, C. Nicolini, *Mat. Sc. Eng. C* 419 (2002)

75.

    (i) X. Hou, L. Sun, M. Xu, L. Wu, J. Shen, Coll. & Surf. B: Biointerfaces (2004) 157

    (ii) X. Hou, L. Sun, M. Xu, L. Wu, J. Shen, *Colloid Surf. B* **33** (3-4) 157 (2004)

76. B. Mechiri, L. Piras, G. Caminati *Bioelectrochemistry* 13 (2004)

77. H. Ringsdorf, B. Schlarb, J. Venzmer *Angew. Chemie. Int. Engl.* 113 (1998)

78.

    (i) D.R. Talham *Chemical Reviews* **104** 5479 (2004)

    (ii) J. T. Culp, J. H. Park, F. Frye, Y. D. Huh, M. W. Meisel, D. R. Talham *Coordination Chem. Rev.* **249** (23) 2642 (2005)

    (iii) J. T. Culp, J. H. Park, M. W. Meisel, D. R. Talham *Polyhedron* **22** (22) 3059 (2003)

    (iv) G. E. Fanucci, M. A. Petruska, M. W. Meisel, D. R. Talham *J. Solid State Chem.* **145** (2) 443 (1999)

    (iv) C. T. Seip, D. R. Talham *Mat. Res. Bull.* **34** (3) 437 (1999)

79. J. T. Culp, J.-H. Park, M.W. Meisel, D. R. Talham *Inorg. Chem.* **42** 2842 (2003)

80. M.A. Petruska, B.C. Watson, M.W. Meisel, D. R. Talham *Chem. Mater.* 14 2011 (2002)

81. S. Kolliopoulou, P. Dimitrakis, P. Normand,H.-L. Zhang, N. Cant, S.D. Evans, S. Paul, C. Pearson, A. Molloy, M.C. Petty, D. Tsoukalas *Microelectronic Engineering* **73** 725 (2004)

82. J.H.Ahn, C.Wang, C.Pearson, M.R.Bryce, M.C.Petty *Appl. Phys. Letts.* **85** 716 (2004)

83. S. Kolliopoulou, P. Dimitrakis, P. Normand, Hao-LiZhang, N. Cant, S. D.Evans, S.Paul, C.Pearson, A. Molloy, M. C. Petty *J. Appl. Phys.* **94** 81 (2003)

84. M. Geard, A. Choubey, B. D. Malhotra, Biosensors Bioelectro 17 345 (2003)

85. G. Cho, D. T. Glatzhofer, B. M. Fung, W-L Yuan, E. A. O'Rear *Langmuir* **16 (10)** 4424 (2000)

86.

    (i) A. Riul Jr., H. C. De Souza, R. R. Malmegrim, D. Santos Jr., A. C. P. L. F. Carvalho, F. J. Fonseca, O. N. Oliveira Jr., L. H. V. Mattoso *Sens. Act. B* **98** 77 (2004)

    (ii) C. E. Borato, A. Riul Jr., M. Ferreira, O.N. Oliveira Jr., L. H. C. Mattoso *Instrum. Sci. Technol.* **32** 21 (2004)

    (iii) M. Ferreira, C. J. L. Constantino, A. Riul Jr., K. Wohnrath, R. F. Aroca, J. A. Glacometti, O. N. Oliveira Jr., L. H. C. Mattoso, *Polymer* **44** 4205 (2003).

87. J. W. Gardner, Microsensors, Wiley, Chichester (1994)

88. P. T. Mosley, A. J. Crocker, Sensor Materials, IOP Publishing, Bristol (1996)

89. M. Haug, K. D. Schierbaum, G. Gaulitz, W. Gopel, *Sens. Act. B* **11** 383 (1993)

90. R. Kasalini, M. Klitzikari, D. wood, M. C. Petty *Sens. Act. B* **56** 37 (1999)

91. T. Sata, *Sens. Actuators B* **23**(1) 63 (1995)

92. T. Okada, K. Hiratani, H. Sugihara, N. Khusizaki *Sens. Actuators, B* **14**(1-3) 563 (1993)

93. S. K. Jeong, B. G. Lee, K. J. Kim, *Bull. Korean Chem. Soc.* **16** (6) 553 (1995)

94. C. A. Lindino, L. O. S. Bulhoes L. O. S. *Anal Chim Acta* **33** (3) (1996)

95. T. Sata *Sens. Actuators B* **23(1)** 63 (1995)

96. G. E. M. Lyons, C. H. Lyons, C. Fitzgerald, P. N. Bartlett, *Electroanal Chem.* **365**(1-2) 29 (1994)

97. M. Pyo, G. Maeder, T. R. Kennedy, J. R. Reynolds *J. Electroanal. Chem.* **368**(1-2) 329 (1994)

98. F. Palmisano, C. Malitesta, D. Centonze, P. G. Zambonin *Anal. Chem.* **67**(13) 2207 (1995)

99. A. Talaie, Z. Boger, J. A. Romagnoli, S. B. Adeloju, Y. J. Yuan *Synth. Met.* **83**(1) 21 (1996)

100. M. Brie, R. Turcu, C. Neamtu, S. Pruneanu *Sens. Actuators, B* **37**(3) 119 (1996)

101. C. Kranz, H. E. Gaub, W. Schuhmann, *Adv. Matter.* **8** 634 (1996)

102. C. de Lacy, J. P. Benjamin, P. evans, N. M. Ratcliffe *Analyst* **121** (6) 793 (1996)

103.

    (i) A. C. Partridge, P. Haris, M. K. Andrew, *Analyst* **121(9)** 1349 (1996)

(ii) J. Janata, M. Josowicz, P. Vany'sek, D. M. DeVaney, *Anal. Chem.* **70** 179R (1998)

104. S. N. Shtykova, T. Y. Rusanovaa, A. V. Kalachb, K. E. Pankina, *Sensors and Actuators B* **1** 497 (2006)

105.

(i) Bansi D. Malhotraa, Rahul Singhala, Asha Chaubeyb, Sandeep K. Sharmac, Ashok Kumarc, *Current Appl. Phys.* **5 (2)** 92 (2005)

(ii) B. D. Malhotra, A. Choubey *Sensor. Actuat. B* **91** 117 (2003)

(iii) V. Saxena, S. Choudhury, S.C. Gadkari, S.K. Gupta, J.V. Yakhmi, *Sensors and Actuators B: Chemical* **107** (1) 277 (2005)

106. M. Penza, E. Mielella, M. B. Alba, A. Quirini, L. Vasanelli, *Sensor and Actuator B: Chemical* **B40** (2-3) 205 (1997)

107. J. A. R. Janssen, O. N. Olivieira Jr. *Colloid Surf. A* **198-200** 45 (2002)

108. M. Penza, E. Mielella, V. I. Anisimkin *Sensors and Actuators B* **47**(1-3 pt 3) 218 (1998)

109. E. Mielella, F. Musio, M. B. Alba, G. Cassano, A. Quirino *Mater. Sc. Eng.* C5(3-4) 255 (1998)

110.

(i) M. Penza, E. Mielella, V. I. Anisimkin, *Ferroelectrics, and Frequency Control* **45(5)** 1125 (1998)

(ii) M. Penzaa, G. Cassanoa, P. Aversaa, F. Antolinia, A. Cusanob, M. Consalesb, M. Giordanoc L. Nicolaisc, *Sensors and Actuators B: Chemical* **111-112** 171 (2005)

111. E. Mielella, M. Penza, Thin Solid Films 327-329 (1998) 694-697

112. E. Mielella, F. Musio, M. B. Alba *Thin Solid Films* **284-285** 908 (1996)

113. R. Çapan, Y. Açıkbaş and M. Evyapan, *Mat. Lett.* In Press, Corrected Proof, Available online 11 May (2006)

114. V. S. Mello, P. Dynarowicz-Latka, A. Dhanabalan, R. F. Bianchi, R. Onmori, J. A. R. Janssen, O. N. Olivieira Jr. *Colloid Surf. A* **198-200** 45 (2005)

115.

(i) A. R. Riul Jr. D. S. Santos Jr. K. Wohnarath, R. Di Tommazo, A. C. P.L. F. Carvalho, F. Fonseca, O. N. Oliveira Jr., D. M. Taylor, L. H. C. Mattoso, *Langmuir* **18** 239 (2002)

(ii) A. R. Riul, R. R. Malmegrim, F. J. Fonseca, L. H. Mattoso, *Artif. Organs* **27**(5) 469 (2003)

116.

(i) V. Zucolotto, A. P.A. Pinto, T. Tumolo, M. L. Moraes, M. S. Baptista, A. R. Jr., A. P. U. Araújo, O. N. Oliveira, Jr. *Biosensors and Bioelectronics* **21** (7) 1320 (2006)

(ii) P. A. Antunes, C. M. Santana, R. F. Aroca, O. N. Oliveira, Jr., C. J. L. Constantino, A. Riul, Jr., *Synthetic Metals* **148** (1) 21 (2005)

(iii) A. Riul Jr., H. C. de Sousa, R. R. Malmegrim, D. S. dos Santos Jr., A. C. P. L. F. Carvalho, F. J. Fonseca, O. N. Oliveira, Jr., L. H. C. Mattoso *Sensors and Actuators B: Chemical* **98** (1) 77 (2004)

117. J. E. G. de Souza, F. L. dos Santos, B. B. Neto, C. G. dos Santos, M. V. B. dos Santos, C. P. de Melo, *Sens. Actuators B.* **88** 246 (2003)

118. W. J. Gardner, P. N. Bertlett *Synthetic Metals* **57** 3665 (1993)

119. Y. Hou, N. J. Renault, C. Martelet, C. Tlili, A. Zhang, J. C. Pernollet, L. Briand, G. Gomila, A. Errachid, J. Samitier, L. Salvagnac, B. Torbiero, P. Temple-Boyer *Langmuir 21*(9) 4058 (2005)

120. O. Worsfold, C. Toma, T. Nishiya, *Biosensora and Bioelectronics* 1505 (2004)

121. X. Guo, J. Chen, L. R. Beuchat, R. E. Brackett *Appl. Environ. Microbiol.* 5248 (2000)

122. E. Racila, D. Euhus, A. J. Weiss, C. Rao, J. McConnell, L. W. Terstappen, J. W. Uhr, Proc. Nat. Ac. Aci, 4589 (1998)

123. J. R. Li, M. Cai, T. F. Chen, L. Jiang *Thin Solid Films* **180 (1-2)** 205 (1989)

124.

(i) M. Rikukawa, M. Nakagawa, N. Nishizawa, K. sanui, N. Ogata *Synthetic Metals* **85** (1-3) 1377 (1997)

(ii) Y. Takeoka, Y. Iguchi, M. Rikukawa, K. Sanui, *Synthetic Metals*, **154** (1-3) 109 (2005)

125. P. Pal, D. Nandi, T. N. Misra, *Thin Solid Films* **239(1)** 138 (1994)

126. J. wang, *Anal Chem* **65(12)** 450R (1993)

127. O. J. Lim, "Design of Novel Intelligent Materials by Incorporating Biomolecules into Electroactive Polymeric Thin Film System (phd) 128p (1993)

128. S. Cosnier, *Biosens. Bioelectron.* **14(5)** 443 (1999)

129. A. Bhaumik, M. Ramakanth, L. K. Brar, A. K. Raychaudhuri, F. Rondelez, D. Chatterji, *Langmuir* **20** 5891 (2004)

130. P. May *Phys. World* **8** 52 (1995)

131. E. H. Rhoderick, "Metal-Semiconductor Contacts", Clarendon Press, Oxford (1978)

132. Y. L. Hua, M. C. Petty, G. G. Roberts, M. M. Ahmed, M. Hanack, M. Rein, Thin Solid Films, 149 (1987) 161

133. A. Aviram, M. A. Ratner *Chem. Phys. Lett.* **29** 277 (1974)

134. A. S. martin, J. R. Sambles, G. J. Ashwell *Phys. Rev. Lett.* **70** 218 (1993)

135. G. S. Ashwell, G. J. Gandolfo *J. Mater. Chem.* **11** 246 (2001)

136.

   (i). G. J. Ashwell, S. H. B. Kelly *Synthetic Metals* **133-134** 641 (2003)

   (ii) R. M. Metzger *Colloids and Surfaces A* **284-285** 2-10 (2006)

   (iii) R. M. Metzger, *Analytica Chimica Acta* **568** (1-2) 146 (2006)

137. E. Racial, D. Euhus, A. J. Weiss, C. Rao, J. McConnell, L. W. Terstappen, J. W. Uhr, *Proc. Nat. Ac. Sci.* 4589 (1998)

138. M. Zhu, G. Liang, T. Cui, K. varahramyan, *Solid State Electron* **47** 1855 (2003)

139. H. Koezuka, A. Tsumura, H. Fuchigami, K. Kuramoto, *Appl. Phys. Lett.* **62** (15) 1794 (1993)

140. J. Matsui, S. Yoshida, T. Mikayama, A. Aoki, T. Miyashita. *Langmuir* **21**(12) 5343 (2005)

141. W. Su, K. Xiao, Y. Chen, Q. Zhao, G. Yu, Y. Liu *Langmuir* **21**(14); 6527 (2005)

142. H. Xu, Y. Wang, G. Yu, W. Xu, Y. Song, D. Zhang, Y. Liu D. Zhu *Chem Phys Lett* **414** 4-6 369 (2005)

143. H. Ohnuki, K. Ikegami, T. Ida, M. Izumi *Colloids Surf. A* **257-258** 381 (2005)

144. H. E. Katz, *J. Mater. Chem.* **7** 369 (1997)

145. M. R. Bryce, M. C. Petty *Nature* **374** 771 (1995)

146. C. D. Dimitrakopoulos, B. K. Furman, T. Graham, S. Hegde, P. Purushothaman, *Synth. Met.* **92** 47 (1998)

147. J. H. Schohn, S. Berg, Ch. Kloc, B. Batlogg *Science* **287** 1022 (2000)

148. Wei Su, Jianzhuang Jiang, Kai Xiao, Yanli Chen, Quanqin Zhao, Gui Yu, Yunqi Liu, *Langmuir* **21** 6527 (2005)

149. M. C. Tanese, G. M. Farinola, B. Pignataro, L. Valli, L. Giotta, S. Conoci, P. Lang, D. Colangiuli, F. Babudri, F. Naso, L. Sabbatini, P. G. Zambonin, L. Torsi *Chem Matter* article in press (2006)

150. A. Lux, S. C. Moratti, X. C. Li, A. C. Grimsdale, J. E. Davis, P. R. Raithby, J. Gruner, F. Cacialli, R. H. Friend, A. B. Holmes, Conjugated Polymers and Oligimers for electroluminescence. American Chemical Society, polymer Priprints, Division of Polymer Chemistry, ACS, Washington, DC, USA 1996 P-202

151. R. H. Friend, R. W. Gymer, A. B. Holmes, J. H. Burroughes, R. N. Marks, C. Taliani, D. D. C. Bradley, D. A. dos. Santos, J. L. Breads, M. Logdlund, W. R. salaneck, *Nature* **397** (6715) 121 (1999)

152. R. j. Mortimer *Electrochim Acta* **44** 2971 (1999)

153. P. Somani, S. Radhakrishnan *Mater. Chem. Phys.* **77** 117 (2002)

154. G. H. Kiss, "Conjugated Conducting Polymers", Springer series in solid state physics, Springer, Berlini, (1992)

155. D. B. Cotts, Z. Reyes, "Electrically conducive Organic Polymers for Advanced Applicatuions". USA, Noyes Data Corporation (1998)

156. B. Scrosati, "Application of Electroactive Polymers", London: Chapman & Halls (1998)

157. A. Bolognesi, G. Bajo, J. Paloheimo, T. Ostergard, H. Stubb, *Adv. Mater.* **9** 121 (1997)

158. V. Cimrov, M. Remmers, D. Neher, G. Wegner *Adv. Mater* **8** 146 (1996)

159. S. E. Burns, N. Pfeffer, J. Gruner, D. Neher, R. H. Friend, *Synth. Mater* **84** 887 (1997)

160. G. Y. Jung, A. Yates, I. D. W. Samuel, M. C. Petty, *Mat. Sci. Eng. C* **14** 1 (2001)

161. K. S. V. Santhanam, N. Gupta, *TRIP* **1** 284 (1993)

162. L. Yuan, B. Li, L. Jiang *Thin Solid Films* **340** (1) 262 (1999)

163. Y. Liu, T. Cui, K. Varahramayan, **47** 811 (2003)

164. J. H. Park, J. M. Ko, O. O. Park, D. W. Kim *J. Power Source* **105** 20 (2002)

165. C. A. Ferreira, S. aeiyach, J. J. Aron, P. C. Lacaze, *Electrochim Acta* **41** 1801 (1996)

166. E. W. H. Jager, E. Smela, O. Inganas, *Sens. Actuators B* **56** 73 (1999)

167. E. W. H. Jager, E. Smela, O. Inganas *Science* **290** (5496) 1540 (2000)

168. V. Saxena, B. D. Malhotra *Current Applied Physics* **3** 293 (2003)

169. M. Aizawa, S. Watarnable, H. Shinihara, H. Sirakawa, "5th International conference on photochemical conversion and storage of solar energy", 1984 p-225

170. M. K. Ram, S. Annapurni, B. D. Malhotra *J. Applied Polymer Science* **60**(3) 407 (1996)

171. E. Punkka. M. F. Rubner *J. electronic Materials* **21**(11) 1057 (1992)

172. T. Cassagneau, T. E. Mallouk, J. H. Fendler *J. Am. Chem. Soc.* **120** 7848 (1998)

173.

   (i) B. Ozturk, G. Behin- Aein, B.N. Flanders; *Langmuir* **21**(10) 4452 (2005)

   (ii) N. Belmen, Y. Cerolan, A. Berman *Cryst. Growth* **5**(2) 439 (2005)

   (iii) M. Wang, K.M. Liechti, Q. Wang, J.M. White *Langmuir* **21**(5) 1848 (2005)

   (iv) P. Bertoncello, A. Notargicomo, C. Niolini *Langmuir* **21**(1) 172 (2005)

   (v) H. Gong, F. Kim, S. Connor, A. G. Somorjai, P. Yang *J. Phys. Chem. B.* **109**(1) 188 (2005)

174.

   (i). M. Matsumoto, H. Tachibana, F. Sato and S. Terrettaz *J. Phys. Chem. B* **101** 5 702 (1997) (ii). M. Matsumoto, F. Sato, H. Tachibana, S. Terrettaz, R. Azumi, T. Nakamura, H. Sakai and M. Abe *Mol. Cryst. Liq. Crys.* **316** 113 (1998)

175. S. Morita, K. Iriyama and Y. Ozaki *J. Phys. Chem. B* **104** 1183 (2000)

176.

(i) B. O. Dabbousi, J. Rodriguez-Viejo, F.V. Mikulec, J. R. Heine, H. mattoussi, R. Ober, K.F. Jensen, M. G. bawendi *J. Phys. Chem. B.* **101** 9463 (1997)

(ii) S. Maenososno, C.D. Duskin, S. Saita, Y. Yamaguchi *Jpn. J. Appl. Phys.* **39** 40006 (2000)

(iii) T. Okamoto, I. Yamaguchi, T. Kobayashi *Opt. Lett.* **25** 372 (2000)

177.

(i) S. Huang, G. Tsutsui, H. Sakaue, S. Shingubara, T. Takahagi *J. Vac. Sci. Technol. B.* **18** 2653 (2000)

(ii) T. Sato, H. Ahmed, D. Brown, B.F.G. Johnson *J. Appl. Phys.* **82** 696 (1997)

178. S. Huang, K. Minami, H. sakue, S. Shingubara and T. Takahagi *Langmuir* **20** 2274 (2004)

179. S. h. Kim, G.Markovich, S. Rezvani, S. H. Choi, K. L. Wang, J. R. Heath *Appl. Phys. Lett.* **74** 2 317 (1999)

180. K. Yano, M. Kyogaku, R. Kuroda, Y. Shimada, Shunichi Shido, H. Matsuda, K. Takimata, O. Albrecht, K. Eguchi and T. Nakagini *Appl. Phys. Lett.* **68**(2) 188 (1996)

181. M. Sastry *J.Mater Chem.* **11** 1711 (2001)

182. N. Pradhan, S. Efrima *J. Phys. Chem. B* **108** 11964 (2004)

183. S. Acharya, S. Efrima *.J. Am. Chem. Soc.***127**(10) 3486 (2005)

184. E. S. Smotkin, C. Lee, A. J. Bard, A. campion, M. A. Fox, T. E. mallouk, S. E. Webber, J. M. White *Chem. Phys. Lett.* **152** 265 (1988)

185. B. O. Dabbousi, C. B. Murray, M. F. Rubner, M. G. bawendi *Chem. Mater.* **6** 216 (1994)

186. A. Dhanabalan, H. Kudrolli, S. S. Major, S. S. Talwar *Solid State Commun.* **99** 859 (1996)

187. A. Dhanabalan, S. S. Talwar, A. Q. Contractor, N. P. Kumar, S. N. Narang, S. S. Major, K. P. Muthe, J. C. Vyas *J. Mater. Sc. Lett.* **18** 603 (1999)

188. I Watanabe, K. Hong, M. F. Rubner, *Langmuir* **6** 116 (1990)

189. M. K. Ram, N. S. Sudershen, B. D. Malhotra *J. Phys. Chem.* **97** 11580 (1993)

190. M. K. Ram, M. Admi, M. Sartore, M. Salerno, S. Paddeu, C. Nicolini *Synth. Met.* **100** 249 (1999)

191. C. Nicolini, V. Erokhin, M. K. Ram *Biosensors and Bioelectronics* **14** 427 (1999)

192. M. K. Ram, S. Carrara, S. Paddeu, C. Nicolini *Langmuir* **13**(10) 89 (1997)

193. M. S. Dresselhaus, G. Dresselhaus, P. C. Eklund, Science of Fullerenes and Carbon Nanotubes", Academic Press: San Diego, CA (1996)

194. H. Dai, J. H. Hafner, D. T. Colbert, R. E. Smelly, *Nature* **384** 147 (1996)

195. S. Wong, E. Joselevich, A. Woolley, C. Heung, C. Lieber, *Nature* **394** 52 (1998)

196. W. A. de Heer, A. Chatelain, D. Ugarte, *Science* **270** 1179 (1995)

197. R. H. Baughman, C. Cui, A. A. Zakhidov, Z. Iqbal, J. N. Barisci, G. M. Spinks, G. G. Wallace, A. Mazzoldi, D. D. Rossi, A. G. Rinzler, O. Jaschinski, S. Roth, M. Kaertesz, *Science* **284** 1340 (1999)

198. S. Tans, A. Verschueren, C. Dekker *Nature* **393** 49 (1998)

199.

(i) X. N. Cao, L. Lin, Y. Y. Zhou, G. Y. Shi, W. Zhang, K. Yamamoto, L. T. Jin, *Talanta* **60** 1063 (2003)

(ii). H. D. Wagner, O. Lurie, Y. Feldman, R. Tenne *Appl. Phys. Lett.* **72** 188 (1998)

200.

(i) V. Bavastrello, S. Carrara, M. K. Ram, C. Nicolini *Langmuir* **20** 969 (2004)

(ii) V. Bavastrello, M. K. Ram, C. Nicolini *Langmuir* **18** 1535 (2002)

201. Yoshiaki Imaizumi, Masahito Kushidab, Yoichiro Arakawa, Fumihito Arai and Toshio Fukuda *Thin Solid Films* **509** (1-2) 160 (2006)

202. T. Jiaoa, Mi. Liua *J. Colloid Interface Sc.* **299** (2) 815 (2006)

203. J-Il. Park, W-R. Lee, S-S. Bae, Y. J. Kim, K-H. Yoo, J. Cheon, S. Kim *J. Phys. Chem. B* **109** 13119 (2005)

204. D. Wang, Y-L. Chang, Z. Liu, H. Dai *J. Am. Chem. Soc.* **9** 127, 11871 (2005)

205. N. Belman, Y. Golan, A. Berman, *Langmuir* (2005)

206. H. Song, F. Kim, S. Connor, G. A. Somorjai, P. Yang *J. Phys. Chem. B* **109** 188 (2005)

207. B. Ozturk, G. Behin-Aein, B. N. Flanders, *Langmuir* **21** 4452 (2005)

208. M. Muramatsu, K. Akatsuka, Y. Ebina, K. Wang, T. Sasaki, T. Ishida, K. Miyake, M. Haga, *Langmuir* **21** 6590 (2005)

209. A. Honciuc, A. Jaiswal, A. Gong, K. Ashworth, C. W. Spangler, I. R. Peterson, L. R. Dalton, R. M. Metzger *J. Phys. Chem. B* **109** (2) 857 (2005)

210. M. Ferreira, C. J. L. Constantino, C. A. Olivati, M. L. Vega, D. T. Balogh, R. F. Aroca, R. M. Faria, O. N. Oliveira Jr. *Langmuir* **19** 8835 (2003)

211. J. Li, V. Janout, D. H. Mccullough, J. T. Hsu, Q. Truong, E. Wilusz, S. L. Regen, *Langmuir*, (2003)
212. J. Zheng, C. A. Constantine, V. K. Rastogi, T. C. Cheng, J. J. DeFrank, R. M. Leblanc *J. Phys. Chem. B* **108** 17238 (2004)
213. Md. N. Islam T. Kato *Langmuir* **21** 10920 (2005)
214. A. S. Alekseev, N. V. Tkachenko, A. Y. Tauber, P. H. Hynninen, R. Osterbacka, H. Stubb, H. Lemmetyinen *Chem. Phys* **275** 243 (2002)
215. N. V. Tkachenko, A. Y. Tauber, V. Vehmanen, A. A. Alekseev, P. H. Hynninen, H. Lemmetyinen, Phytochlorin-fullerene dyad: Photoinduced electron transfer in solutions and solid LB films, pp. 161-171 in "Fullerenes 2000 Volume 8. Electrochemistry and photochemistry". Eds. S. Fukuzumi, F. D'Souza, D. M. Guldi, Electrochemistry Society, 2000
216. O. Oliveira Jr. *Analytical Chemistry* **73** 3674 (2001)
217. K. Wohnrath, C. J. L. Constantino, P. A. Antunes, P. M. dos Santos, A. A. Batista, R. F. Aroca, O.N. Oliveira, Jr.; *J.Phys. Chem. B* **109** 4959 (2005)
218. T. Del Caño, P.J.G. Goulet, N.P.W. Pieczonka, R.F. Aroca, and J.A. De Saja *Synthetic Metals* **148** 31 (2005)